THE MYSTERY OF
NUMBERS

THE MYSTERY OF

NUMBERS

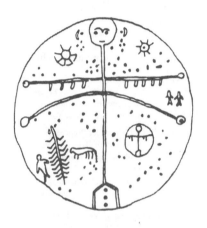

ANNEMARIE SCHIMMEL

OXFORD UNIVERSITY PRESS
NEW YORK   OXFORD

Oxford University Press

Oxford   New York   Toronto
Delhi   Bombay   Calcutta   Madras   Karachi
Kuala Lumpur   Singapore   Hong Kong   Tokyo
Nairobi   Dar es Salaam   Cape Town
Melbourne   Auckland   Madrid

and associated companies in
Berlin   Ibadan

Originally published in Germany as *Das Mysterium der Zahl* by Eugen Diederichs
Verlag, Munich. Copyright © 1984 by Eugen Diederichs Verlag
First published as Oxford cloth in 1993 by Oxford University Press, Inc.,
200 Madison Avenue, New York, New York 10016

First issued as an Oxford University Press paperback, 1994
Oxford is a registered trademark of Oxford University Press

Library of Congress Cataloging-in-Publication Data
Endres, Franz Carl, 1878-1954.
[Mysterium der Zahl, English]
The mystery of numbers / Annemarie Schimmel.
p.cm. Translation of: Das Mysterium der Zahl
/ Franz Carl Endres, Annemarie Schimmel.
Includes bibliographical references and index.
ISBN 0-19-506303-1
ISBN 0-19-508919-7 (PBK.)
1. Symbolism of numbers.
I. Schimmel, Annemarie. II. Title.
BF1623.P9E55 1992
133.3'35-dc20 90-22456

2 4 6 8 10 9 7 5 3 1

Printed in the United States of America

PREFACE

When publisher Ulf Diederichs decided to produce a book on numerology and number symbolism, he first thought of issuing a new edition of one of the numerous publications in this field that had appeared in Germany, or a translation of a comprehensive work written in English. As I had contributed some articles on numbers to encyclopedias, he asked me to help him with the selection. After carefully going through the material available to us, I found Franz Carl Endres's *Das Mysterium der Zahl* most fitting for our purposes. I planned to add examples from my own field, Islamic studies, which understandably seemed most interesting to me. While working on the additions, however, I discovered that a good number of very important books and articles on the broader subject of numbers had appeared in the decades since the last edition of Endres's book in 1951.

Most of the modern publications dealt with medieval allegoresis of numbers, among them V. C. Hopper's classic work and the interesting studies of Heinz Meyer. A history of *Zahlwort und Ziffer* was offered by Karl Menninger in a comprehensive work, which was translated into English as *Number Words and Number Symbols* (1969), and Willi Hartner devoted a very extensive and fascinating article to the different number systems in the world. Furthermore, some of the theories presented by F. C. Endres, like so many others in the science of religion, have become outdated or have been somewhat modified by more recent scholarship.

Given all of these circumstances, we decided to rework the book completely. The reworking is based on the structure of the original text and makes use of some of its material, but in the process of incorporating as much new information as possible, an independent work came into being. While this was first published under Endres's and my names in acknowledgment of his basic contribution, the name of Endres has now been dropped from the title page of the English edition with the permission of his only surviving daughter and heir.

Needless to say, it is impossible to attempt completeness in a work like this, and every reader will probably be able to add details from his or her own field of interest and expertise about one number or another, its application, or its "magic" character. We hope that the bibliography will help those who are interested in special topics to pursue them further.

The field of numerology and number magic has fascinated humanity throughout the millennia. The sun and moon, these signs in the great book of nature that serve to measure human life, have always made people feel that numbers are something very special. Not only do they circumscribe and determine space and time in abstract formulas, but they are also part of a mysterious system of relationships with the stars and other natural phenomena. Earlier generations usually considered these phenomena to be connected in turn with, or to represent, spirits, deities, or demons. To know a number and the powers inherent in it made it possible for mortals to use this power to secure the help of the relevant spirits, to perform witchcraft, or to make their prayers more efficient by repeating certain formulas in prescribed numbers. The knowledge of the secret meaning of numbers is reflected both in folklore and in high literature; it can be discovered in medieval architecture just as it is captured in the music that is thought to manifest the harmony of the spheres.

This book mainly concentrates upon the civilizations of antiquity and the religious traditions of Judaism, Christianity, and Islam. However, we have tried to incorporate Indian and Chinese traditions as far as possible and to cast a glance at the Mayas. It would be a fascinating task to explain in detail the number symbolism of the pre-Columbian Americans, who developed such admirable astronomical systems, or to dwell upon numbers in the civilizations of the American Indians or among African tribes.

Number symbolism is extremely variegated, and yet, amazing similarities in interpreting numbers can be discovered among different cultures. We do hope that in the end the readers will not be "at sixes and sevens" but will have learnt something about the context from which number symbolism and number magic developed. At least they will be able—or so we hope—to understand why the traffic light has three phases, why many people select particular "lucky" numbers for their license plates (often with seven or multiples of seven), and why so many people suffer from the ailment with the frightening name of Triskaidekaphobia, that is, "fear of thirteen."

We thank all those who have contributed remarks from their own fields of expertise or from the native traditions. My special thanks go to Miriam Rosen, who edited the text.

Bonn, Germany   A. S.
September 1991

CONTENTS

INTRODUCTION

A LITTLE DICTIONARY OF NUMBERS

INTRODUCTION

Numbers and Number Systems

> The mathematical spirit is a primordial human property that reveals itself wherever human beings live or material vestiges of former life exist.

Thus writes Willi Hartner in a fundamental study about numbers and number systems. In support of this argument, he cites the example of the Stone Age artist, who, he contends, did not count and did not know anything of mathematical relations, but rather, relied exclusively on his mathematical instinct. It is this instinct, then, that has been abstracted and fettered into geometrical forms. In the course of time, it helped to develop the concept of numbers and then the numbers themselves, so that finally the manifold manifestations of being in space and time could be put in order by abstract numbers. As Karl Menninger has shown in another fine book, this ordering process could find a multiplicity of possible expressions, a multiplicity that is likely to amaze people like us who are used to seeing everything within the framework of our inherited decimal system and the Arabic numerals. Even the Anglo-Saxon system of weights and measures is not easy to grasp for someone coming from the German or French tradition.

Number systems are built according to different rhythms. One experiences this when trying to understand the binary system underlying the computer—the bases of which were developed by Leibniz as early as 1697. And even though the decimal system seems to be the most widespread, one has to admit that other systems are equally important. Among them, the sexagesimal system in ancient Babylon deserves special mention: there, after the first unit of 10, the second, higher unit is formed by 60. This division survives in the seconds, minutes, and hours of the day, as well as in the degrees of the circle.

In many cases, computing systems and numerals are simply derived from the 5 or 10 fingers: the Roman numerals, for example, tell us by their shape that they originated in finger signs. Excluding the thumb yields another way of counting, up to the number 4, and indeed, in many civilizations a new section of counting begins after 4. Alternately, the 10 fingers could be combined with the toes to form a vigesimal system, and there is a strong possibility that such a system was known to Celts, Basques, and other peoples in the north and west of Europe. Even today, in French, 80 is called *quatre-vingts*, or 4 × 20. The English *score* similarly reminds us of an ancient vigesimal system.

One must not confuse the fundamental finger numbers with the technique of finger counting, which was formerly so highly developed that one could count up to 100 by using different configurations of the fingers. The units and tens were formed with the fingers of the left hand, the hundreds with the right hand. Medieval exegetes of the Bible, especially the Venerable Bede, referred to this way of counting, which held special meaning for them: the thumb and index finger of the right hand, closed into a circle, for example, indicated 100, and became as it were a symbol of closed eternity. European merchants used this system until the late Middle Ages, and merchants in the Middle East still use it with incomprehensible swiftness. Also used in the East, especially in China, is the abacus, the calculating device with little counters that slide along rods to perform the most complicated operations at almost lightning speed. One should also not forget that our term *calculus* is derived from *calcule*, "pebble," a reference to counting with pebbles.

Every civilization had its own signs for numbers. One may think of the knotted *quipus* of the Incas or of the tally (German *Kerbholz*) on which debts were incised with different kinds of cuts. The German expression *Etwas auf dem Kerbholz haben*, "to have something on one's tally," in the sense that one has committed some sins or illegal actions, reflects the latter way of counting. In ancient Egypt, the numbers were pictorial, while the Phoenicians and later the Romans used comparatively primitive forms of numbers. To perform more complicated mathematical operations, other systems proved more practical, such as that in which the letters of the alphabet also represented numbers. This method is found in early Greek and still exists in Hebrew and Arabic. In the latter case, the Arabic alphabet follows the old Semitic sequence of letters, called *abjad*, and since each letter has a twofold meaning, one can easily develop relations between names, meaningful words, and numbers (as has been done in the Cabala for centuries). The number 666 in the Book of Revelation is a model case: numerous interpreters have found in it the names of persons who seemed to personify the "Beast" of their own time. In the Islamic tradition, the art of producing elegant chronograms was highly developed, and in later times the title of a book might be used to record the date of its completion: the Persian title of the book *Bagh u bahar* (Garden and spring), for example, shows by its numerical value $(2 + 1 + 1000 + 6 + 2 + 5 + 1 + 2 + 200)$ that it was composed in the year 1216 of the Islamic era (1801/2 c.e.). One could similarly give the date of a person's death by applying a fitting word or sentence, and this art has been practiced with particular skill in the eastern Islamic countries.

Willi Hartner considers the Chinese number system to be superior to this kind of letter-numerals that are so widespread in magic and mysticism, but even more advanced, according to him, are the highly abstract Sumerian and Babylonian number systems. In fact, ancient Mesopotamia was a place where astronomy and mathematics developed from early times, and we owe to the Mesopotamians many of the meanings of certain numbers as we use them today (for example, the sacredness of 7, the importance of 60). But according to Hartner, the highest development of the numerals is found in the system of the Maya, whose astronomical calculations involved enormous numbers, and which was of an amazing exactitude. Indeed, their very ancient calendar, based on 65 synodic revolutions of the planet Venus around the sun is more exact than any other calendric system.

As for our own "Arabic" numerals, their Indian origin can be easily recognized from the fact that they are written from left to right even when used in Arabic script, which runs from right to left. The Indian system, which the Arabs adopted soon after the emergence of Islam, includes the zero, which permits very complicated mathematical operations. Indian sources call it *shunya*, "emptiness," that is, an emptiness that fills the lines between the numbers and thus makes it easy to distinguish the position of a number in terms of units, tens, and so on. The Indian sources use this expression as early as the sixth century A.D.; in the Middle East the nine Indian numbers are first found in a Syrian book dated 662 C.E. Long before the West knew anything of these practices, Arab scholars were using them to compose mathematical works. Muhammad ibn Musa al-Khwarizmi's book *Hisab al-jabr wa'l-muqabala* (The book of restitution and equation) was written soon after 800 and was translated into Latin by Robert of Chester about 1143. This book, the earliest introduction of Arabic numerals, not only provided the West with the concept of "algebra" (*al-jabr*), but also with the term *algorithm*, which is nothing but the misspelled name of its author, Khwarizmi. However, Arabic numbers were received and accepted in Europe quite slowly. It was the ingenious Pisan scholar Leonardo Fibonacci (d. 1250) and John of Sacrobosco who strove to introduce the Arabic numbers and to explain their infinite possibilities. Around 1240, Menninger tells us, a French Franciscan friar, Alexander de Villa Dei, was so excited by the new mathematical discoveries that he wrote the *Carmen de algorismo*, a poem of 244 verses about the new way of computation which, he thought, had been invented by an Indian king by the name of Algor.

The zero, which had been unknown in previous numerical systems, caused much confusion, as becomes clear from the history of its very name. From its Arabic name, *sifr*, were derived *cifra*, *chiffre*, and the German *Ziffer* on the one hand, *zero* on the other. This zero, which in itself does not mean anything but imparts to the numbers preceding and following it their proper rank, was regarded as late as the fifteenth century as *umbre et encombre*, "dark and encumbered," and its German name, *Null*, is derived from the idea that it is *nulla figura*, not a "real" figure. But the zero was not restricted to India and from there the Islamic world and finally Europe: the Maya, and perhaps before them the Olmec, had known zero

Left: The Maya sign for zero is the empty oyster, *xok*. The word *xok* means basically any round, curved thing and more precisely, "hollow," taken to define the object and its character. The Mayas' ingenious invention was to use zero also to define the value of a number's position. *Right:* the sign for 20.

The so-called head variants of the Maya numbers are glyphs that represent 1 to 19 and zero through profile heads of deities. From 1 to 13, the deities are different; 14 to 19 show the same figures as 4 to 9 with the difference that a bone is placed over the chin (an attribute of death). Compare 5 and 15, 8 and 18. For the head variant of zero, a hand is placed on the chin.

completely independent of the Indian discovery and, as it seems, earlier than the Indians. In the Maya vigesimal system, zero followed the number 19. To note down this system the Maya used either dot-line combinations or glyphics in the shape of heads.

Just as numbers and number systems are not the same or even similar all over the world so we should not presume that all civilizations use the same way of counting or computing. In the introduction to his book on number symbolism, F. C. Endres tells how, at the beginning of this century, he tried a mathematical experiment in some faraway Turkish villages: "On this occasion I asked a boy to count some apples which I had placed on the ground. He tried with the help of his fingers but could not get farther than 5. Between 5 and 10 he frequently made mistakes, and when I put more than 10 apples before him and told him to count them, he simply said that there were many, but could not name a specific number." When the same author cast pebbles in a brook, the boy's sequence in counting never went farther than 3 or 4: while counting something visible in space, he could at least make use of his fingers and see what was being counted, but to count something in time is much more problematic, since one has to remember how often an act is repeated, either exactly or in a similar way. It seems that a two- or threefold repetition of sounds, shouts, or rhythmical units is generally recognized, but other groups and sequences appear—at least to us—rather difficult to count. One can experience this when trying to follow the complicated rhythmical patterns of Armenian or, even more, Indian music; one usually loses track very soon and is not able to continue counting correctly. And then there is the case of some African tribes that can barely "count" according to our understanding of the term, yet, in a way that looks uncanny to us, they are able to recognize even large amounts of objects, so that they know immediately if only a single animal is missing in a large herd.

In some cultures, numerical words are related to the counted objects, and the expression used for 6 long items may be completely different from that used for 6 cows or 6 plants. Such classifications are rather common. The Papua discern up to 20 such groups of numbers, each of which is related to the counted object. Even in our tradition one finds traces of such counting—woven material is measured according to the yard (in German *Elle*), height in feet, depth of water according to fathom (German *Faden*), the speed of ships in knots (*Knoten*, nautical mile). Number-words are especially common with the different species of animals, such as the pack of dogs (German *Meute*), the string of horses (German *Koppel*), the flock of sheep (German *Herde*), and the chain of partridge (*Kette*.) Such concepts are still very much alive in the hunters' idiom, as one can understand from J. Lipton's delightful book *An Exaltation of Larks* (1977). In measuring certain household items or foodstuff, one finds such German expressions for eggs as a *Schock* (threescore), a *Stiege* (score), and *Mandel* (fifteen). The army is divided into companies, squadrons and so on—all words that carry a true numerical value. And when the German speak of 2 *Stück Vieh*, two "pieces" of cattle, or the English of a "brace of partridge," they resemble the Per-

sians, who possess a whole set of counting words that cannot
be translated but are used to refer to animals of different
species: *yak zinjir fil* is *one* elephant, not, as a literal translation would have it, a "chain of elephants"!

None of these expressions has a mystical meaning or is
connected with magic, and yet, already in early civilizations
one feels that numbers are a reality having as it were a magnetic power-field around them: as Levi-Bruhl formulates it,
they can "work." Or, as it was claimed in ancient India, the
number is "Brahma-natured," meaning that it is similar to
the Divine. Indeed, in certain texts of ancient India numbers
are worshiped: "Hail to the One, hail to the 2 . . . , hail to
the 100. . . ." Such feelings about the special character of
numbers have been passed on from generation to generation,
and even in our apparently rather sober and unmystical
number system that can be reduced to the 10 fingers, the 4
phases of the moon, and the 12 months of the year, some
mysterious overtones have always been maintained. Thus,
numbers have been attributed with special, secret powers that
make them fitting for magical conjurations and, of course, for
astrological prognostications. Even the "high" religions recognize the religious importance of certain numbers and their
mystical character, not only in the Middle Ages but to this
day. In magic, where practitioners invoke certain formulas to
influence events for their own benefit or to the detriment of
others, the correct use of numbers plays an immense role, for
each number is seen in its power-field and in its cosmic connections and thus, the use of the right number, along with the
correct number of repetitions and magic formulas, of purifica-
tions and circumambulations, is considered to be absolutely
decisive for the success of the magic act.

THE HERITAGE OF THE PYTHAGOREANS

In our cultural sphere, that is, the Jewish, Christian, and Islamic one, the interest in numbers and their specialties is mainly based on foundations laid by the Pythagoreans. Born in the sixth century B.C.E. on the island of Samos, Pythagoras emigrated in 532 B.C.E. to Kroton in southern Italy to escape Polycrates the tyrant. He may have lived for some time in the East, particularly in Egypt, where he would have learned something of the mathematical knowledge of the ancient East.

Every school child learns the Pythagorean theorem of the right triangle, according to which the square of the hypotenuse equals the sum of the squares of the two shorter sides, but just as this formula is part and parcel of our mathematical knowledge, other ideas of the master and his disciples have influenced religious, literary, and even magical works. The centerpiece of Pythagoras's thought is the idea of order: musical order, mathematical order, the order of the cosmos, and finally the ethical and social order. He is said to have discovered that the intervals of the musical scale correspond to the relative lengths of the vibrating strings, which he expressed by the ratios 1:2, 2:3, and 3:4. Thus, the first four integers were established, and the Pythagoreans have never ceased to emphasize their importance.

Just as musical harmony could be expressed in mathematical formulas by measuring the ratios of the strings, so the essence of everything seemed to be expressible in numbers. Observation of the regular movements in the sky led to the concept of a beautifully ordered harmony of the spheres. The evolution of the world was paralleled by that of numbers: unity came into existence from the void and the limit; out of

Pythagoras discovers the relations between the order of numbers and the frequency of sounds. He is shown experimenting with bells, water-filled glasses, strings, and pipes of different sizes. Opposite him his Hebrew counterpart, Jubal, is shown making instruments. Woodcuts from F. Gaffurio, *Theoria musica* (Milan, 1492).

the One the number appears, and out of the number comes the whole heaven, the entire universe. As Bell writes for the Pythagoreans, "the cosmos is isomorphic with pure mathematics" and "everything in the universe is measurable by common whole numbers." In their system there was no room for irrational numbers; thus, the discovery, ascribed to Hippasus, that the ratio between the side and the diagonal of the rectangle cannot be expressed in integers shattered the Pythagorean worldview. It is said that the discovery of a fifth solid body, the pentagon-dodecahedron (a three-dimensional pentagon, with twelve faces) shocked them even more.

Perhaps the most influential disciple of Pythagoras was Philolaos, who was active about 475 B.C.E. and apparently developed many ideas concerning the meaning of numbers which remained alive in later number mysticism.

The Pythagoreans were especially fascinated by the difference between the odd and even numbers. It has been speculated that the ratios between simple harmonies (1:2, 3:4) might be at the source of this interest. The Pythagoreans went so far as to divide everything in the universe into two categories: the odd numbers belong to the right side, which is associated with the limited, the masculine, the resting, the straight, with light and goodness, and, in terms of geometry, with the square, while the even numbers belong to the sphere of the infinite, the unlimited (as they are infinitely divisible), the manifold, the left side, the female, the moving, the crooked, darkness, evil, and, in geometrical terms, the rectangle.

This contrast between the one and the many, as expressed by odd and even numbers, is emphasized later, especially in mysticism, with its goal of undivided, absolute unity. Odd numbers therefore have played an important role in popular belief, and even in theological speculations. For Plato, all even numbers were of ill omen, and Hopper states correctly: "As if the feminine numbers were not already sufficiently in disfavor, the stigma of infinity is attached to them, apparently by analogy to the line." Virgil claims: "Numera deus impare gaudet" (The deity is pleased with the odd number), and the same idea is taken up in the Islamic tradition, where it is said: "Verily God is an odd number (*witr*, that is, "One") and loves the odd numbers." Shakespeare, too, states: "There is divinity in odd numbers" (*The Merry Wives of Windsor*, v.i.2).

This predilection for odd numbers has led to the custom that ritual acts, prayers, incantations and so on are repeated an odd number of times. One performs acts of magic 3 or 7 times and repeats a prayer or the concluding "amen" thrice. In earlier times, physicians and medicine men used to give their patients pills in odd numbers. Magic knots too had to be tied in odd numbers. The Talmud offers numerous examples of the use of odd numbers and the avoidance of even ones, and the Muslim tradition states that the Prophet Muhammad broke his fast with an odd number of dates. When performing witchcraft or black magic, an odd number of persons should be present, and even today it is the custom, in Europe at least, to send someone bouquets containing an odd number of flowers (with the exception of a dozen).

Among the other concepts that the Pythagoreans introduced to mathematics is that of the perfect number, one whose components, when added up, produce the number itself. The first of these is 6 (1 + 2 + 3); the next one is 28 (1 + 2 + 4 + 7 + 14); to date 23 such perfect numbers have been discovered, the last in 1971. Recently, there has even been an attempt to explain the mysterious title of Hugo von Hofmannsthal's tale *The Story of the 672nd Night* by the fact that 672 is a double perfect number, but that seems farfetched.

The Pythagoreans further related numbers to geometrical forms; 3,6,10,15 are triangular numbers; 1,4,9,16,25 are square numbers (i.e., $1^2, 2^2, 3^2, 4^2, 5^2$). The dot belongs to 1, the line to 2, the space, which appears first in the triangle, to 3, and the body, surrounded by 4 spaces, to 4. The most perfect number in the Pythagorean system was 10, since it is the sum of the first 4 integers (1 + 2 + 3 + 4) and could be represented as an equilateral triangle. Thus, multiplicity again became unity in the 10. For this reason the Pythagoreans strove to discover 10 heavenly bodies in order to fit them into their system of cosmic order, and in the absence of a tenth one, they invented it. Aristotle (384–322 B.C.E.) wrote somewhat critically about Pythagorean number mysticism in the first book of his *Metaphysics*, where he states that, completely submerged in mathematics, they assume that their number principles are the principles of everything existing:

> As in mathematics, the numbers are by nature the first thing, the Pythagoreans thought to recognize in numbers many likenesses of what exists and what will be, such as the elements of fire, air, earth, and water; they furthermore found the qualities and relations of musical notes in numbers and thus considered the elements of numbers to be the elements of everything existing, as everything seemed to be formed according to numbers, which were regarded as the first thing in all of nature, and they believed the entire vault of heaven to be harmony and numbers. One of the manifestations of numbers was supposed to be justice, another one the soul or the intellect; other forms of manifestations were time and occasions, and thus everything that exists at all. And they collected the correspondences between numbers and harmonies on the one hand and the qualities and parts of the sky and the whole world on the other hand and compared them. And if there was something missing, an artificial glue had to help to produce relations everywhere in the system. For example, as the number 10 seems for them the most perfect thing and appears furthermore to embrace the whole realm of numbers, as a result there must also be 10 bodies circling in the sky as stars. But as there are only 9 visible ones, they invented a special tenth body, an invisible counter-earth.

The number singled out by Aristotle as pointing to justice is the 4, because it is the product of equal factors, that is, the first square number. For the Pythagorean thinkers such equations proved the objective measures of harmony and beauty that they strove to discover.

Gnosis and Cabala

With the continuing quest for measures of life and for an all-embracing harmony, even Plato, otherwise somewhat critical of the Pythagoreans, accepted that the numbers contained certain keys for solving the mysteries of nature. Pythagorean and Platonic ideas were carried over in Neoplatonism and the gnostic systems and gave rise to a number mysticism that can briefly be summarized as follows:

1. Numbers influence the character of things that are ordered by them.
2. Thus, the number becomes a mediator between the Divine and the created world.
3. It follows that if one performs operations with numbers, these operations also work upon the things connected with the numbers used.

In this way, every number develops a special character, a mystique of its own, and a special metaphysical meaning.

Plotinus, whose Neoplatonic system deeply influenced the mystical trends in the three Abrahamic religions, Judaism, Christianity, and Islam, remarked: "Numbers exist before the objects described by them. The variety of sense objects merely recalls to the soul the notion of number." Continuing this line of thought, Philo of Alexandria combined ideas from the Old

Testament and the Pythagorean tradition and thus created the basis for the biblical exegesis of the Middle Ages, which is heavily determined by number mysticism. The most important development of the Pythagorean tradition in the medieval world, however, is the Jewish Cabala, which is based upon a highly complicated number mysticism, whereby the primordial One divides itself into 10 *sefirot* (from *safar*, number), which are mysteriously connected with each other and work together, with the 22 letters of the Hebrew alphabet serving as "bridges" between them. The highest *sefirah* is *keter* (Crown), out of which *hokhmah* (Wisdom) and *binah* (Intelligence) branch out. The fourth *sefirah* can be called *hesed* (Love) or *gedullah* (Greatness); the fifth one is *gevurah* (Justice); the sixth one *tiferet* (Beauty), and the seventh *netsah* (Triumph). Added to them are *hod* (Splendor) as the eighth, *yesod* (Fundament) as the ninth, and finally *malkhut* (Kingdom or Reality). This last *sefirah* can also be equated with the *Shekhinah* that lives in the exile of this world. Out of the 10 *sefirot*, which are perhaps best called *logoi* or primordial ideas, comes the world of the first divine emanation, *atsilut*. There are three more worlds, which also depend upon the 10 *sefirot*: the world of creation and the heavenly spheres, *beriah*; the world of the figurations of creatures connected with the heavenly realm, like spirits and angels, *yetsirah*; and finally the world of matter, *asiyah*.

Since the Hebrew letters also serve as numbers, the figure of the *sefirot* and its derivations lead to fascinating relations between the different parts of the world. The vast field of cabalistic hermeneutics, masterfully described by Gershom Scholem, is permeated by number mysticism.

ISLAMIC MYSTICISM

It seems that the proto-Ismaili group of the Ikhwan as-Safa, the Brethren of Purity, in Basra, who composed their encyclopedic treatises in the tenth century C.E., were the first to use Neoplatonic and Pythagorean ideas extensively. According to them, Pythagoras was a sage from Harran, and they believed that the prophet sent to the Sabians in Harran was Enoch, that is, Hermes Trismegistos, who, they held, was particularly well versed in number mysticism. For the Brethren of Purity numerology was a way to understand the principle of unity that underlies everything. It is a science that is above nature and yet is the root of all other sciences. Thus the relation of God to the world, or of Pure Being to existence, is equated with the relation of the 1 to the other numbers. Although the Ikhwan as-Safa did not create a complicated system of numerology comparable to that of the Cabala, they were well aware of the importance of numbers as seen in music and in the order of the cosmos alike. The numbers 7 and 12 play a particularly important role. Seven, the number of the planets, is the first complete number, since it can be obtained from 3 + 4, 2 + 5, and 1 + 6—that is, it is the sum of the numbers on the opposite sides of a die. Nine, meanwhile, is important as the number of the spheres and as the first odd number from which the square root can be drawn, while 12, the number of the zodiac, is a combination of either 3 × 4, or 5 + 7. Finally, 28 is the most perfect number of all: it corresponds to the number of lunar mansions and as such is especially connected with Islam as it also corresponds to the number of letters in the Arabic alphabet. Among mystically minded Muslims, the possibility of interchange between letters and numbers has led to highly sophisticated operations in the realms of quranic exegesis, prognostication, and frequently poetry, especially in the clever use of chronograms.

MEDIEVAL AND BAROQUE NUMBER SYMBOLISM

Medieval Christianity shared the same tradition that had been carried on among the gnostic sects. As Isidore of Seville wrote around 600 C.E., "Tolle numerum omnibus rebus et omnia pereunt" (Take from all things their numbers, and all shall perish). Number symbolism, combined with astrological ideas, permeated medieval thought, and the Church used it profusely. As Heinz Meyer observes, "Does the Bible not state that all things are ordered in measure and number and weight (Wisdom 11:21)? Nothing in the universe could be without order, and thus, as Augustine holds, numbers are the form of divine wisdom, present in the world, which can be recognized by the human spirit." Biblical interpretation by means of allegories based on numbers has survived at least to the nineteenth century.

The so-called science of arithmology developed in the first centuries of our era as a kind of philosophy of the powers and virtues of particular integers. Among mathematical works, the *Introductio arithmetica* (Introduction to Arithmetic) by Nicomachus of Gerasa (*ca.* 100 C.E.), which was popularized by Boethius, exerted great influence during the Middle Ages. Besides this useful introduction, the author also composed a theology of numbers, which, unfortunately, is extant only in fragments. Another important medieval text was the *Mathesis* of Firmicus Maternus (*ca.* 346), which was devoted to astrology in particular. Even greater was the influence of Martianus Capella from Carthage, who lived in the fifth century. In his work *De nuptiis philologiae et Mercurii* (On the Marriage of Philology and Mercury), he surrounded the bride with 7 bridesmaids symbolizing the 7 liberal arts, and the seventh book, "Arithmetica" (which relies largely on Nicomachus) shows how the first ten integers are connected with Greek deities. Beginning with the 1, the monad, which corresponds to Jupiter, Capella proceeds through the 2 of Juno (feminine, number of separation and reunion) to 10, the number of the two-headed god Janus. Among other standard works of medieval number mysticism were *De Numero* (On the Number) by Hrabanus Maurus and the *Liber numerorum qui in sanctis scripturis occurent* (The Book of Numbers That Occur in the Holy Scriptures) by Isidore of Seville.

Medieval scholars were so convinced that numbers were extremely important and effective that they sought to arrange their writings in meaningful combinations of numbers. Augustine's *City of God* is a prime example of this tendency. The 22 sections correspond to the 22 letters of the Hebrew alphabet and are divided into 2 times 5 refutations, which is an expression of the tenfold "Thou shalt not . . ." of the Law, and into 3 times 4 positive teachings, which correspond to both the 12 apostles and the Trinity as proclaimed in the 4 gospels: $3 \times 4 = 12$. Even more refined is the elaborate number symbolism in Dante's *Divine Comedy*, which is based on the 3 of the Trinity.

A similar meaningful division is also known from medieval Muslim theological works, the best example of which is al-Ghazzali's *Revivification of the Sciences of Religion* with its 40 chapters. Forty is the number of preparation and is used

The thirty-third canto of the *Purgatorio* and the thirty-third canto from *Paradiso* in Dante's *Divine Comedy* (1311–1321). Woodcuts from the Milan edition of 1864. The *Divine Comedy* consists of 3 parts, each containing 33 cantos; an introductory canto serves to complete the "perfect" number 100.

in this work to lead the reader through the works of the Law and the acts of mystical love to the final chapter, which is devoted to the meeting with the Lord at the moment of death. It seems important that the central, twentieth, chapter is devoted to the central figure in Islam, the prophet Muhammad.

Contemporary with Ghazzali in the Christian world was Hugo of St. Victor and his circle. Speaking of the different means by which the Scripture can be understood through number mysticism, Hugo claims that there are 9 different ways of recognizing the significance of numbers. The first way is by order of position: 1, for example, is connected with Unity and is the first number, the principle of all things. Alternately, one can look at their composition: 2, for example, can be divided and points to the transitory. Another meaning can be discovered through extension: since 7 follows 6, it means rest after work. Numbers can also have a meaning according to their disposition: thus, 10 has one dimension and points to the right faith, while 100 expands in width and thus points to the amplitude of charity, and 1000 rises in height and can therefore be taken as an expression of the height of hope. One may also look at the numbers in connection with their use in the decimal system, in which case 10 means perfection. Another way to find a special meaning in a number is through multiplication: 12 is universal because it is the product of the corporeal 4 and the spiritual 3. Likewise, the number of its parts can be taken into consideration: 6, as is well known, is a perfect number because it is the sum of its integral components. It is also possible to look at the units that make up a number: 3 points to the Trinity, consisting of 3 units. And finally one may use exaggeration to understand why 7, under certain circumstances, grows into 77.

Clearly, such techniques made almost every possible interpretation permissible. Similarly, one could use the position of a certain number in the decimal system: thus 11 could be taken as a positive power advancing beyond 10, but more frequently it was seen as a negative number that transgresses the closed system of the 10.

Meyer, whose argument we are following here, has shown very convincingly that virtually all the things whose numbers are specified in the Bible have been transformed into symbols that then gain a special value of their own by their use in the liturgy. The ecclesiastical year and the liturgical service are largely formed by the use of such numbers, which are interpreted allegorically. Even the ordinal numbers of the Psalms were explained according to numerological rules and were then used in exegesis. Still another field of activity was medieval sacred architecture, the numerological foundations of which were transmitted in the workshops of architects and masons with an admirable mathematical and technical knowledge.

In recent years the construction of medieval literary works according to numerological rules has been studied rather frequently, sometimes with conflicting results; Meyer's book seems to me to be the best guide through this labyrinth.

Whatever one says, it is certain that number mysticism played a very important role in the European Middle Ages and Renaissance. This is evident from the many scholarly works appearing on this topic in the sixteenth and seventeenth centuries, beginning with Giorgio's *Harmonia mundi* (1525), in which numerology is used as a kind of super-science through which all other scholarly disciplines can be unified. Nearly a century later, Petrus Bungus composed an enormous encyclopedia, *De numerorum mysteriis* (1583, 2d edition 1618), in which he claimed that without the knowledge of numerology it is simply impossible to understand why there are only 4 elements and only 7 planets. Bungus's work, which was recently reprinted with a very useful introduction by Ulrich Ernst, offers a marvelous survey of the use of numbers, rich in quotations from classical, medieval European, and even Arab thinkers and astronomers. Plato, whom Bungus considers the "head and leader" of number mysticism, appears in this work as an "atticized Moses," for it was believed that the entire Egyptian, meaning ancient Oriental, numerological wisdom had been known to Moses before it reached Greece. In Germany, Agrippa von Nettesheim's *De occulta philosophia* (1533) is a veritable compendium of numerology. One century later, Athanasius Kircher composed his *Arithmologia, sive de abditis numerorum mysteria* which can be considered the most comprehensive description of numerology; published in Rome in 1665, it became well known all over Europe.

The use of numerological principles did not end with the Middle Ages: the Rosicrucians, for example, developed their own numerology, and their writings are strongly permeated by number mysticism. Renaissance writers drew on the same principles. Suffice it to mention Milton's *Doctrina christiana*

(Christian Doctrine), the first volume of which is divided into 33 chapters according to the age of Jesus at the time of the Crucifixion, and the second volume of which contains 17 chapters, alluding to the Ten Commandments and the 7 gifts of the Holy Spirit (and thus the round number of 50 is reached without any difficulty). Similarly, Pico della Mirandola's *Heptaphis* offers 7 interpretations of the 7 days of creation in 7 books with 7 chapters each. And when John Donne, in his poem "A Valediction: forbidding mourning," compares himself to the compass which "ends where it began," the beauty of this image is enhanced by the fact that the poem consists of 36 lines, corresponding to the 360 degrees of the circle.

One may call such formulas learned games, but it should not be forgotten that the belief in a mathematically ordered world, in the *harmonia mundi*, even led an ingenious astronomer like Kepler to make some of his discoveries, for he had, as Hartner states, the unshakable conviction that there existed a harmony between human beings, the earth, and the cosmos—a harmony ruled by number.

It is natural that this *harmonia mundi* is also expressed in the musical harmony that was one of the roots of Pythagorean number mysticism. From at least the third century onward, medieval musical theory knew the *musica coelestis* (heavenly music), and in the sixth century, Cassiodorus wrote, in truly Pythagorean style, that *musica est disciplina quae de numeris loquitur*—music speaks of numbers. Thus medieval and, even more outspokenly, Renaissance composers turned to the sacred and mysterious numbers to use them in the technique of the canon, in the number of voices to be employed, and in the continuo. The 3, be it in the form of the triad or in a chorus of 3 voices, was seen as especially related to the Trinity, while the number 7, as in 7 voices, was often used in compositions in honor of the Virgin Mary. Such allegorical use of numbers, widespread in the seventeenth and early eighteenth centuries, was especially common with J. S. Bach, whose late works have been called "largely musical mathematic" because he exhausted the different possibilities to utilize meaningful numbers to their limits. A good example is the sevenfold repetition of the *Credo* in the B Minor Mass.

Even in German classicism and romanticism the knowledge of the deeper meaning of numbers was apparently well established, as is evident from Goethe's numerous allusions to the mystery of this or that number. An early, incomplete poem, *Die Geheimnisse* (The Mysteries), reflects the number symbolism of the Rosicrucians. Schiller too used the traditional meanings of numbers, especially in his drama *Wallenstein*, and Novalis was convinced "that in nature a wondrous mysticism of number works, and also in history."

SUPERSTITIONS

The ancient belief in the order of numbers may have led, as in the case of Kepler, to scientific discoveries, but much more frequently it led to magical manipulations, and such a belief in the power of number mysticism has survived to this day. Indeed, on the popular level, it seems to have increased. In the spring of 1984 an American catalog of books and journals that landed in my mailbox proposed publications exclusively devoted to numerology, the vibrations of numbers, the discovery of one's lucky number, and so on. It seems that these practices have not changed, or lost their attraction, from the days of classical antiquity. Bell's book *Numerology* offers perhaps the most trenchant condemnation of such superstitious games, against which Franz Carl Endres was also always outspoken. There are too many things that can be manipulated, and the skillful use of any name and any date can lead to the hoped-for result. Operations involving the total of the digits of dates and of the number of names are especially liable to produce the most astounding results. One may also think in this connection of results that have been achieved in a "perfectly scientific" way. Thus, at the end of World War I, a German scholar, Oskar Fischer, tried in his book *Auferstehungshoffnung in Zahlen* (Hope of Resurrection in Numbers) to interpret certain combinations of numbers in the Old Testament by means of statistics and then to draw far-reaching conclusions for the history of Israel and early Christianity. There are also the fascinating books by E. G. McClain, who tries to reconstruct the mystical cosmic mountain and to establish its relation to the Kaaba in Mecca by means of numerological principles and who, apart from this all-pervasive numerological aspect of his work, makes very interesting remarks about certain aspects of Islam. Indeed, we may also remember the attempt of a pious Muslim to prove, with the help of the computer, that the Quran is built completely upon the number 19.

Even such well-intended but not very meaningful attempts to organize historical and mythical data according to numerological principles cannot disturb our natural pleasure in the harmony of numbers or the innate feeling of some people that certain number constellations repeat themselves time and again in their lives. But this is a border zone in which both tradition and psychology play important roles. We are barely aware of how many aspects of our daily routine or language are formed by a numerical rhythm in which the 3 plays a predominant role. Such ternary rhythms, to give only a few examples, appear in such trivialities of everyday life as the 3 phases of the traffic light or the triple cheer at a birthday party. Certain numbers have been used over and again to convey a specific atmosphere in a literary work, and often words and expressions may have come unconsciously to the poets, and even to the scholars, when they continue their diction in triadic steps or in a fourfold rhythmical structure, when they use distinct metrical devices for the contents of their verse or, like Goethe, create a *Trilogie der Leidenschaft* (Trilogy of Passion) or, in the Persianate world, organize their epic works in a *khamsa,* or quintet. The same holds true for the artist who strives to come as close as possible to the golden section. There are apparently certain subconscious structures that lead later interpreters and exegetes of a literary or artistic work to see intended number mysticism while the author may not have been aware of this at all.

And should we blame the Pythagoreans for considering that 2 and all even numbers are feminine, when modern biology proves that the sex-distinctive form in the chromosome has the pattern xy for the male, but the pattern xx, that is, an even number, for the female?

NUMBER GAMES AND MAGIC SQUARES

The quantity of number games is almost without end, and many of them are very amusing, but the majority are purely arithmetic and never applied to mystical or magical use. Many of these games can reveal their beauty only through the use of Arabic numerals. In such games, a very special place belongs to the 9. At an early point, for example, mathematicians discovered that all multiples of 9 also have 9 as the sum of their digits: $4 \times 9 = 36$, sum of the digits 9; $7 \times 9 = 63$, sum of the digits again 9, and so on. Similarly, 5 and 6, raised to their powers, always produce sums that end with 5 or 6 respectively: $5^2 = 25$, $5^3 = 125$, and so on. These are the so-called circular numbers. One can also build little number trees by multiplying certain numbers, and here again the 9 has a special place:

$$1 \times 9 + 2 \ = 11$$
$$12 \times 9 + 3 \ = 111$$
$$123 \times 9 + 4 \ = 1111$$
$$1234 \times 9 + 5 \ = 11111$$
$$12345 \times 9 + 6 \ = 111111$$
$$123456 \times 9 + 7 \ = 1111111$$
$$1234567 \times 9 + 8 \ = 11111111$$
$$12345678 \times 9 + 9 \ = 111111111$$
$$123456789 \times 9 + 10 = 1111111111$$

An even more interesting tree begins with $1 \times 8 + 1 = 9$ and treats the following numbers the same way as in the first number tree, with a last line that looks like this:

$$123456789 \times 8 + 9 = 987654321$$

And is it not surprising that when multiplying the number 142,857 by 2, 3, 4, or 6, one always gets a result that consists of the same numerals, although in a different sequence?

Number games go back to the Middle Ages and were developed further during the Renaissance, but it seems that they have become even more popular in the last few decades. Between 1925 and 1970, more than 300 new publications about all sorts of number games, including paradoxes and alphamatics (letters substituted for numbers) were published, and today they are found not only in specialists' journals but in the Sunday issues of many newspapers.

A medieval number game also led to the discovery of an

important numerical sequence, which is known as the Fibonacci series. The famous mathematician Leonardo Fibonacci of Pisa (*ca.* 1170–*ca.* 1250), who was, among other things, responsible for the acceptance of the Arabic numerals among Western scholars, wondered how a couple of rabbits would increase in number if they produced two young ones each month, and these young rabbits, from the second month of their lives, in turn produced a couple of young ones every month, and so forth.

Month	1	2	3	4	5	6	7	8	9	10	11	12 . . .
Couples	1	1	2	3	5	8	13	21	34	55	89	144 . . .

In the resulting series, 1,1,2,3,5,8,13,21 . . . , each number is the sum of the two preceding ones. This appears in numerous forms in nature, such as fir cones and the petals of certain flowers, where it is called phyllotaxis. A prime model for this sequence is the disposition of the chambers in the nautilus shell.

Magic squares are another important category of number games. The discovery of the first magic square is recorded in the following anecdote: In ancient China a just and wise monarch ruled between 2205 and 2198 B.C.E.; this was the emperor Yu, known for his great wisdom and care. Kungfutse tells that the emperor had once been occupied with building a dam on the Yellow River to stop the floods. While he was sitting on the bank of the river, immersed in thought, a divine turtle named Hi appeared to him. On the turtle's back there was a figure with number signs which, transcribed into modern numerals, looked like this:

4	9	2
3	5	7
8	1	6

One immediately sees that the square is grouped around the number 5, which was highly esteemed in ancient China. All the horizontal and vertical lines produce the sum of 15, as do the diagonals. The even numbers are placed in the corners, the odd ones between them.

This square is most common in the Islamic tradition, for it is believed that it contains the 9 letters that were revealed to Adam, that is, the first 9 letters of the Arabic alphabet in the old Semitic sequence (which is used to this day when letters are taken as numbers). As for the even numbers in the corners, they are read according to their numerical value as *buduh*, and this word, sometimes interpreted as the name of a spirit, often appears on walls to protect the building, or on amulets worn around the neck or on the upper arm.

The numbers around the central 5 could be arranged in different sequences, and each of the transformations was thought to be connected with one of the 4 elements. Thus, the original is the fire square, which is related to water, and another form of the square is related to earth. The squares are used in magic according to these properties.

6	1	8
7	5	3
2	9	4

Fire Square

2	7	6
9	5	1
4	3	8

Earth Square

One can produce such squares with every arithmetical sequence of first order on perfect-square fields, such as 16, 25, 36, 49, etc. These squares, each with its specific contents, were then assigned to the different planets. The original number, with the sum of 45 on 9 small squares, is connected with Saturn (and here one should remember that 45 is the numerical value of the Arabic name of Saturn, *zuhal* = 7 + 8 + 30). The Jupiter square consists of 16 fields; the Mars square, of

Jupiter amulets. The Jupiter square consists of 4 lines adding up to 34 each, for a total of 136. If this configuration is engraved on a silver tablet during the time that the planet Jupiter is ruling it is supposed to produce wealth, peace, and harmony.

25; the sun square, of 36; the Venus square, of 49; the Mercury square, of 64; and the moon square, of no less than 81 fields.

It seems that these squares, which were very common in the Islamic tradition, reached the West comparatively late, probably in the fifteenth century. One of the most famous examples of their use is the design in Dürer's etching *Melencolia I* (Melancholy). Behind the angel, or winged genius, and in the midst of various symbolic instruments, one sees a square consisting of 36 fields, that is, a true Jupiter square.

Mars amulets. The tablet consists of 5 lines with 13 in the center; the sum of each line is 65, for a total of 325. If engraved on an iron plate or a sword when the planet Mars is in the ruling position, this amulet is supposed to bring success in lawsuits and victory over the owner's enemies.

The central numbers of the lowest row read together give 1514, the year in which the etching was made. In addition, the square has the Jupiter number, 34, in all the horizontal, vertical, and diagonal lines; 34 is also the sum of the number of the corners of the large square (16 + 13 + 4 + 1) and the number of the smaller central square (10 + 11 + 6 + 7), and the rest of the numbers result in 68 = 2 × 34.

The Arabs ascribed great power to magic squares. Certain squares were shown to a woman in labor and then placed over her womb to facilitate the birth. One also sees them embroidered or written on the shirts of warriors, mainly in the Turkish and Indian areas. (Such a shirt had to be made by 40 innocent virgins in order to work.) Squares could also be formed from divine names, or from the mysterious letters at the beginning of quranic chapters (*suras*), especially the beginning of chapter 19, *k h y ᶜs*, with the numerical value of 20, 5, 10, 70, and 80. In some cases the results of such attempts are not true magic squares since the sums of the horizontals and the verticals are not equal, but quite often a perfect square was constructed from a divine name. Thus, the name *matin*, "the Firm One," was used to protect and help lame children. Its square looks like this, formed from *m* (40), *t* (400), *y* (10), and *n* (500):

50	10	400	40
40	50	10	400
400	40	50	10
10	400	40	50

The divine name *hafiz*, "the Preserver," was built into an irregular square of *h* (8), *f* (80), *y* (10), *z* (900) in order to obtain the sum of 998 for each line:

900	10	80	8
7	81	9	901
12	902	6	78
79	5	903	11

Numerous medieval Arab authorities describe the construction of such squares and enumerate the rules to be followed. The most famous, as well as the most comprehensive compendium of esoteric knowledge is al-Buni's thirteenth-century *Shams al-maʿarif* (The sun of knowledge). Squares, as well as cabalistic letter and number mysticism, were often used for prognostication. Starting with the numerical value of a name, a date, or a place, out of which one drew the sum of its digits, which was then multiplied by an important number such as 7, or from which one subtracted certain other numbers, one could understand whether a marriage was likely to be happy, an ailing person to recover, or a traveler to reach home safely.

The art of interpreting letters and numbers, gematria, is known from Babylonian inscriptions of the time of Sargon II (723–705 B.C.E.). It could be used in infinite varieties, exchanging words or letters of the same numerical value, making permutations on the words, or creating new words from each letter of a word or a name. Arabic and Persian writers liked to play with the root letters ʿ r sh, which could produce the words *shrʿ* (law), *shʿr* (poetry), and *ʿarsh* (throne). Such games were known earlier in antiquity, a good example being *The Alexander Romance* by Pseudo-Callisthenes, and they were dear to the Jewish tradition, where Rabbi Eliezer declared in the first century C.E. that gematria was the twenty-ninth of the 32 ways to interpret a text, thus giving this art an official place in Jewish spirituality. Gematria also plays an important role in Sufism. Even in the modern West, gematrical games are by no means restricted to a few cabalistically-minded people. Indeed, the equation *Bonaparte = 82 = Bourbon* is well-known—historical figures are connected by means of gematria and, as Franz Dornseiff states in his informative scholarly work, "It is ridiculous how often it works correctly." Thus, Goethe, Schiller, and Shakespeare, among others, were interpreted in this way.

Another application of this kind of letter and number

magic is the interpretation of dreams. An Arabic story tells that a woman had dreamed that a cat (*sinnaur*) had put its head into her husband's belly to take something away. The interpreter correctly stated that someone had stolen 316 dirhams from her husband since 316 is the numerical value of the word *sinnaur*. Such applications of number mysticism and magic squares remain common to this day, as one understands from a book like Erich Bischoff's *Mystic und Magie der Zahlen* (Mysticism and Magic of Numbers), which teaches the reader how to make nativity squares for famous people. Indeed, this is the branch of mathematical games that has been most used, or misused, for magic purposes. Besides the comprehensive literature about the traditional expressions of such techniques in the Islamic East, an increasing literature concerning the modern use of such magic numbers is being published today.

We, however, are concerned with the pleasure derived from playing with numbers, a pleasure one finds, for instance, in folk songs and children's rhymes where the first 10 integers are frequently used for more or less meaningful purposes. That is also true for counting-out rhymes, such as this little Yiddish song about 10 brothers with its decreasing numbers:

> *Zehn Brider sennen mir gewesen,*
> *Hobn mir gehandelt mit Wein—*
> *einer is fun uns geschtorben,*
> *Sennen mir geblieben nain. . . .*

> (We were ten brothers, selling wine—
> one of us died, then we were nine.)

There is a German counterpart about the *zehn kleine Negerlein* (10 little Negroes), and who would not think of the English rhymes about the twelfth day of Christmas, when:

> my true love gave to me 12 lords a-leaping, 11 ladies dancing, 10 pipers piping, 9 drummers drumming, 8 maids a-milking, 7 swans a-

swimming, 6 geese a-laying, 5 gold rings, 4 calling birds, 3 French
hens, 2 turtle doves, and a partridge in a pear tree!

Such verses are often used in riddles and fairytales but can
also be found in higher literature: the following Persian verse
is only one of the numerous examples of this art in the Persianate world:

> The *ten* friends from the *nine* spheres and the *eight* paradises
> and the *seven* stars from the *six* directions have written this letter:
> Among the *five* senses and *four* elements and *three* souls, God
> has not created in *both* worlds a *single* idol like you!

(For the concepts see the chapters in "A Little Dictionary of
Numbers.")

Books are often arranged in ascending or descending
length of their chapters: the Buddhist *Angutara nikaya* offers
the Buddha's words in ascending length, while in the Quran
the chapters are arranged according to decreasing length, preceded only by the short "Fatiha" (Opening) and closed with
two prayers for protection. Religious contents can alternate
with profane groups, as in the English folk song "Green grow
the rushes - 0 . . ." where the 12 apostles, the 10 Commandments and the 4 Gospel-makers are combined with the 7 stars
in the sky, the 6 proud walkers, and so on, while the end
points to the truth:

> One is one and all alone
> And evermore shall be so.

The best example for enclosing the entire basic religious
knowledge into such a series of numbers is probably the recitation from the Jewish Passover Haggadah, which begins:

> Who knoweth one? I, saith Israel, know one: one is the Eternal, who
> is above heaven and earth.
>
> Who knoweth two? I, saith Israel, know two: there are two

tablets of the covenant; but one is the Eternal who is above heaven and earth

and culminates in the final statement:

Who knoweth thirteen? I, saith Israel, know thirteen: there are thirteen divine attributes, twelve tribes, eleven stars, ten command-ments, nine months preceding childbirth, eight days preceding cir-cumcision, seven days in the week, six book of the Mishnah, five books of the Law, four matrons, three patriarchs, two tables of the covenant, but one is the Eternal who is above heaven and earth.

A LITTLE
DICTIONARY
OF NUMBERS

THE NUMBER OF
THE PRIMORDIAL BEING

○ ○ ○ ○ ○ ○ ○ **1**

> The One permeates every number. It is the measure
> common to all numbers. It contains all numbers united
> in itself but excludes any multiplicity. The One is al-
> ways the same and unchangeable, that is why it has
> itself as a product when multiplied by itself. Although
> without part, it is divisable. However, by division it is
> divided not into parts but rather into new units. None
> of these units, however, is larger or smaller than the
> whole unit, and every smallest part of it is again itself
> in its wholeness.

This description of the qualities of the mystical One, written
by the medieval German mystic Agrippa of Nettesheim
around 1500, cannot be considered to be mathematically correct, but it serves to show the importance of the 1 in religious
traditions.

One, geometrically represented by the point, was not regarded by the Pythagoreans and the thinkers under their influence as a real number because a number is an aggregate
composed of units, as Euclid holds. Kobel wrote in 1537:
"From this you understand that 1 is not a number but is a
producer (or 'mother'), the beginning and foundation of all
other numbers." Since 1 is the first originator of numbers,

and even though it is odd, it was considered both male and female, although it was somewhat closer to the male principle. When added to a male number, it results in a female number, and vice versa: $3 + 1 = 4$, $4 + 1 = 5$.

One became the symbol of the primordial One, the divine without a second, the nonpolarized existence. It comprises relation, entirety, and unity and rests in itself but stands behind all created existence. Real unity is inconceivable, however, for as soon as a self thinks about itself, there is a duality: the observer and the observed. Polarity is essential to recognition: whatever is qualified with attributes can only be recognized because of polarity. Large and small, high and deep, sour and sweet—all such qualities are relative to an ordering system. The divine, however, is beyond polarity; it is absolute being, unrelated to any ordering system. For this reason, the mystics have tried to approach the final unity beyond the manifestations, the *deus absconditus* beyond the *deus revelatus,* by using negations, and have spoken, as in the Upanishads, of *neti neti,* "not not," or in the Cabala, of the Ein Sof (literally "without end"), which then reveals itself in descending layers of manifestations. In Islam, the problem for the serious believer posed itself in the question: Can a human being really pronounce the words of the profession of faith, 'There is no deity save God,' without committing the major sin of *shirk,* that is, to associate something with the One? The most radical Sufis have formulated the opinion that only after the seeker's complete annihilation in the divine, God Himself speaks through His mouth the profession of His own unity.

Such mystical thoughts were well known in the ancient Indian tradition, which claims that the primordial and all-pervasive principle is the One without a second; similar words are used by Plotinus (d. 270), the most influential Neoplatonic thinker in late antiquity, whose ideas provided a basis for the development of mysticism in the Jewish, Christian, and Muslim traditions. For him, the One God is beyond all forms, because forms always express multiplicity. As every multiplicity is a multiplicity of unities, it presupposes unity in itself. And since God is the root and presupposition of everything, He is also absolute Unity.

Ancient Chinese religion has expressed similar ideas about 1, which represents the All, the Perfect, the Absolute beyond all polarity.

One is an ideal symbol of the divine because the divine is spirit and as such has nothing to do with material qualities, which are bound to appear in multiplicity. Thus, 1 has no contrast, and even the negative principle that seemingly opposes the deity is, in the end, annulled or integrated into the Unity. As 1, God is the Absolute One as well as the One that is unique in its being.

The Indian thinkers of the Upanishads were in search of the unity behind the varied manifestations that were taken as mere appearances, as ways of action, indeed, as deceptive phantasms before the One, or as a colorful play that veils the essential Unity. "Manifold, that is how poets call the One that is only one. One is the fire that flares up in ever so many forms, One is the sun, radiating upon the world," as an Indian sage described it. The German poet-orientalist Rückert has expressed such ideas in his didactic poem *Die Weisheit des Brahmanen* (The Wisdom of the Brahmin), taking over in the last line a formulation from chapter 112 of the Quran, the chapter that is the shortest expression of God's unity:

> So wahr als aus der Eins die Zahlenreihe fließt,
> So wahr aus *einem* Keim des Baumes Krone sprießt,
> So wahr erkennest du, daß der ist einzig Einer,
> Aus welchem alles ist, und gleich ihm ewig keiner.
>
> (As truly as the chain of numbers emerges from the 1
> As truly as the tree's foliated crown grows out of a single seed
> As truly you recognize that He is One and unique

He, from whom everything emerges and to whom nothing is
Equal nor eternal like Him.)

To attain this unity that lies hidden behind the manifold
manifestations and to achieve identity with the One has always been the goal of mysticism, as expressed in its classical
form in the Upanishads: *aham brahmasmi*, "I am Brahma."
Besides this inclusive mystical monotheism, however, there
exists another type of monotheism, which has been called
"prophetic" or exclusive. This is a religious form most clearly
seen in Judaism and Islam (very strict monotheists do not
consider Christianity, with its doctrine of the Trinity, to be
truly monotheistic). This type of monotheism is likely to have
developed out of henotheism, which is the adoration of one
special god who is superior to the other deities, and to whom
believers turn in their worship more than to others. The biblical Yahweh, however, teaches his prophets that he, the One, is
the only God, a jealous God who does not tolerate the worship
of anyone else. Similarly, the Islamic profession of faith
claims that "There is no deity save God," and "deity" can be
interpreted as anything that diverts humanity from the absolute surrender (*islam*) to this one creator, sustainer, and
judge. To be sure, in both Judaism and Islam a more mystical,
more inclusive type of piety developed, which can be studied
in the Cabala on the one hand and in the Sufi tradition on the
other (particularly in the school of Ibn 'Arabi), both of which
show Neoplatonic trends. At that point the sober and clearcut
statement that "there is no deity save God" can be expressed
by the formula, "there is nothing existent save God," and the
creation out of nothing by God's creative word is interpreted
as a mysterious doctrine of emanations from the One.

Prophetic monotheism is not restricted to the Semitic
peoples. In the fourteenth century B.C.E. the Egyptian king
Amenhotep IV proclaimed a monotheistic religion that had

the sun as the one central deity. However, the opinion that *urmonotheism* preceded all other religious forms cannot be maintained, in spite of how much material its major defender, Pater Wilhelm Schmidt, collected to prove his point.

Christianity, although viewed with suspicion by the stern monotheistic religions of Judaism and Islam, has in fact ascribed the greatest importance to the One. Suffice it to read, in Paul's *Epistle to the Ephesians* (4:5): "One lord, one faith, one baptism, one God, father of all". And as the Venerable Bede has claimed, the one temple in Jerusalem pointed to the one eternal home of all believers.

The German Protestant mystic Valentin Weigel has expressed the mystery of 1 as the number of the deity in a beautiful saying: "The One is a conclusion and concept of all numbers, 2,3,5,10,100,1000. Therefore you can say that 1 is all numbers *complicite*—rolled together—and 2,3,5,10,100, 1000 is nothing but unfolding. I shall compare God to the first number and the creatures to the other numbers because God is one . . . and because the creature in itself is twofold or has two aspects, one to itself, and one to God." And the seventeenth-century Catholic mystic Angelus Silesius sings:

> Gleich wie die Einheit ist in einer jeden Zahl
> So ist auch Gott der Ein' in Dingen überall.
>
> (Just as unity is in every number,
> thus God the one is everywhere in everything.)

For the Muslim mystics the numerical value of 1 for the letter *alif*, the first letter of the Arabic alphabet, as well as the first letter of the name of Allāh, God, offered a wonderful possibility for puns and allusions on every literary level. Is it not enough to know the first letter of the alphabet and its numerical value, since it contains all wisdom and knowledge in itself? The person who has come to know the One God does not need anything more.

POLARITY AND DIVISION

> Die Zwei ist Zweifel, Zwist, ist Zwietracht, Zwiespalt,
>   Zwitter,
> Die Zwei is Zwillingsfrucht am Zweige, süß and bitter.
>
> (Two is doubt, disunion, discord, dissension, her-
>   maphrodite,
> Two is the twin fruit on the twig, both sweet and
>   bitter.)

Rückert invented this ingenious wordplay in his didactic epic *The Wisdom of the Brahmin* to allude to many negative characteristics of the number 2. In religious traditions, 2 means disunion, the falling apart of the absolute divine unity, and is therefore the number connected with the world of creation: "creature is twofold in itself," as Valentin Weigel said in the sixteenth century.

In every aspect of life one observes the central position of dichotomies, dyads, and dual structures. These, however, need not point to a purely negative discord. For the very possibility of discussion, of addressing someone other than one's self, contains a tension between I and Thou, a tension that can be fruitful as well as fatal. The line, the geometrical expression of 2, separates as it unites.

Many languages possess the grammatical form of the dual, which was the subject of an article by the German thinker Wilhelm von Humboldt as early as 1828. This dual, which expresses the relation between two individuals but not between the I and a larger group, is found in antiquity and remnants still exist in certain German dialects, notably in some areas of Westphalia and Bavaria-Austria. Here, as in some Slavic languages, it occurs particularly in personal pronouns. In a different linguistic tradition, the Semitic one, Arabic has completely retained dual forms, while modern Hebrew has discarded them, although they appear throughout the Old Testament.

The confrontation between I and Thou contains, by its very nature, an opposition, and such an opposition becomes even more evident when the human I is confronted with the absolute, unique divine Thou. For as we saw above, it is impossible to think of anything truly opposed to the divine One. Thus, 2 becomes a number of contradiction and antithesis and, logically, of the non-divine. Since it produces discord, it is rarely used in magic.

When reading these deliberations one must not think of the mathematical equation $1 + 1 = 2$, for from the esoteric and mystical viewpoint, only the One that cannot be repeated or doubled exists. One god plus another god makes 2 deities, which, then, no longer correspond to the ideal of the eternal, unique One. For this reason, in religious and magic thought, 2 has always been the symbol of the confrontation of two relative units rather than divine ones. It was again Agrippa of Nettesheim who gave a fine description of the religious peculiarities of the number 2: it is "the number of man, who is called 'another,'" and it is the number of the lesser world. It is also the number of sex, and of evil, for, as all medieval exegetes of the Bible emphasize, in the story of creation in Genesis, the formula "and it was good" is lacking on the

Adam and Eve, or the unification of the male and female principles in the signs of sun and moon. Woodcut, sixteenth century.

second day. Thus it was believed that "fearful goblins" and "mischievous spirits that molest travelers" are under the reign of 2. The fact that "the second" is, in Scandinavian languages, *andra*, "the other," fits well with Agrippa's formulation, which places mortals as "another" against God.

Two came into existence only with creation, because without the polarity it expresses, material life would not exist. As the electric current needs a positive and a negative pole and animal life continues through inbreathing and outbreathing and by the heartbeat with its systole and diastole, 2 is connected with all the manifestations in the world of creatures.

As Goethe, whose whole work shows his awareness of the
mystery of polarity, writes, using a purely Islamic image:

> Im Atemholen sind zweierlei Gnaden . . .
> Du danke Gott, wenn er dich presst
> Und danke' ihm, wenn er dich wieder entlässt.

> (There is twofold grace in breathing . . .
> you should thank God when he presses you
> and equally when he releases you.)

This saying takes up the alternating steps of *qabd*, "pressure,"
"depression," and *bast*, "elation," "expansion", as they were
experienced by the Sufis of the Islamic world and expressed in
ever new images. Just as fear cannot be conceived without the
complementary feeling of hope, pressure and release inevita-
bly belong together. For God the One, in order to become
known, manifests himself in what Rudolf Otto has called the
mysterium tremendum and the *mysterium fascinans*. But this
was known in the Islamic tradition many centuries ago, for
the Muslims hold that God manifests himself through his
beauty and kindness, *jamal*, and his majesty and wrath, *jalal*,
both of which point to his unfathomable, unique perfection in
which all the opposites come together.

Like the Cabalistic mystic, Islamic Sufis have discovered
an allusion to the created world in the second letter of the
alphabet, *b*, which in both Hebrew and Arabic has the numerical value of 2. And just as the Bible begins with *b'reshit*, "In
the beginning . . . ," the first words of the Quran are
Bismillāh, "In the name of God . . . ," so that in both cases
the first letter of the holy book is the letter of creation, the *b*.

A beautiful symbol of the duality that appears through
creation was invented by the great Persian mystical poet Jalaladdin Rumi, who compares God's creative word *kun* (written in Arabic *KN*) with a twisted rope of 2 threads (which in

Left: Yin and Yang, in China the natural powers of the female (the dark) and the male (the light). Both have emerged from the primordial One (*T'ai chi*). The union of the polarized powers produced the 5 "changing powers" or elements out of which the "10,000 things" came into existence. *Right:* Representation of the dyadic number sequence, the binary system as discovered by Leibniz. Reverse side of a sketch for a memorial medal described by Leibniz in a letter to Prince Rudolf August of Braunschweig, dated 2 January 1697: "I have sketched on it light and darkness, or according to human representation God's spirit above the water. . . . And this is all the more fitting as the empty abyss and gloomy waste belong to zero and nothing, while the spirit of God with his light belongs to the omnipotent One."

English is called *twine,* in German *Zwirn,* both words derived from the root "two"). This twisted yarn appears in all manifestations of creation but dupes only the ignorant, who are led to believe in multiplicity, while the wise know that the world of unity is hidden behind the apparent contrasts.

Perhaps the most ingenious way to show the fundamental polarity on which life rests is found in the yin and yang of the Chinese religion, by which active and passive, male and female, begetter and begotten, fire and water, day and night, and whatever complementary relations exist are expressed. These relations are extremely subtle and appear in both cosmic and human relations. Thus the highest celestial being and the highest ruler appear in the yang principle, while the moon, the water, and the empress are related to the yin principle. Yin and yang are omnipresent and inseparable, for it is impossible to make either one absolute: the mention of "woman" includes involuntarily the idea of "man," just as "health" presupposes the existence of "illness." (These examples are taken from the Persian mystic Rumi's work, not from Chinese, but they express the yang-yin relationship with perfect lucidity.) In another area of Chinese thought, the 2 types of small sticks used in the oracles of the *I Ching* bear witness to the highly developed art of divination based on psychological concepts (and a dyadic system long before the invention of the computer).

Under the strong impression of the reality of evil, some religions developed a dualistic worldview. The best-known, and still extant example is Zoroastrianism, the ancient Iranian religion, with its contrasting pair of Ahura Mazda, the god of light and goodness, and Ahriman, the dark, evil principle. For the Zoroastrians, everything in the world belongs to one of these two powers. It was, however, only with later gnostic systems, especially Manicheism, that the good principle was tied up exclusively with the spiritual while the evil was connected with everything material, hence, the soul had to strive to escape the evil, material prison of this world and this body.

Insofar as the number 2 expresses the breaking up of the primordial unity of being and points to the alienation of creation from its ultimate source, mystically oriented religions emphasize its negative aspects. The "prophetic" religions, on the other hand, have discovered a positive value in this very tension between the One and the many, between creature and creator, for their goal is not so much the final unification of the creature with the divine One (like the raindrop that loses itself in the ocean), but rather the creative dialogue, the awareness of the relation between I and Thou in prayer. This very separation and its expression in longing and in unending quest, resulting from the knowledge that the separation between the eternal One and the one created in time cannot be overcome, has inspired many poets in East and West alike. The German poet Rudolf Alexander Schroeder (d. 1962) sings, for example:

> Ich möchte Dir nimmer so nah sein,
> Daß ich mich nach Dir nicht sehnte. . .
>
> (I would not like to be so close to Thee
> that I would cease longing for Thee. . .)

Similarly, among Muslim thinkers, Schroeder's elder contemporary Muhammad Iqbal (d. 1938) expresses in ever-new images the importance of longing, which he sees as humanity's truly creative spark, unlike the eternal quietude of mystical union, which he finds dangerous and unproductive. However, the longing for a reunion of what has been separated by the very act of creation permeates most religions. Thus Goethe, addressing his beloved, hopes that "a second word 'Be!' shall not separate us for a second time" (Und ein zweites Wort "Es werde!" / Trenn uns nicht zum zweitenmal. . . .)

In many cultures sexual union is construed as a means of overcoming this polar opposition of male and female. Thus, in the tantric tradition of the Hindus and Buddhists, one tries to attain to the experience of absolute unity through sexual practices. In Hinduism, even the god needs his *shakti*, his feminine power, and even in the deepest and generally rather abstract mysticophilosophical texts such as the Upanishads, the experience of the *unio mystica* is compared to the loving union of man and woman: "Just as someone embraced by a beloved woman no longer knows what is inside and outside," thus the mystic, embraced by absolute unity, no longer recognizes any duality between god and creature. In the Jewish tradition the Cabalists likewise speak of the *hieros gamos*, the sacred union of male and female powers as symbolized in the pairs of begetting and receiving potencies in the divine unity.

This is particularly evident in the combination of the ninth *sefira*, Yesod, with the tenth one, the Shekhinah.

It is therefore not astonishing that in many religious traditions primordial man is imagined as androgynous (in German, *Zwitter*, derived from the root "two"). Thus there is a whole group of Hindu statues that represent the deity as half male and half female. According to Tacitus, even in the ancient Teutonic religion there was a primeval androgynous being called Twisto, who was regarded as the father of Manno, the "male" human. Such a state, seen as one of higher perfection, is even alluded to in an agraphon of Jesus that is quoted by Clement of Alexandria: "Then Salome asked Jesus: 'When shall thy kingdom come?' And the Lord spoke: 'When two have become one, when the male has become female and when there is no female any more.'" In other words, the kingdom of God will return once duality has been overcome. For duality appeared, as the Torah and later the Cabala hold, because man ate from the tree of knowledge and thus became aware of the existence of good and evil and of life and death, as well as of sexuality. "And thus you have endangered yourself," says the *Zohar*, the central text of the Cabala, "because everything that originated from the tree of knowledge carries in it duality."

The Christian church also had negative interpretations of the number 2, seen as a deviation from unity, from the first good. Is it not written in the Bible that 2 unclean animals from each race were taken into the ark? Furthermore, 2 is the number associated with heretics, those who, in the words of Gregory the Great, are *duplex cor*, having two hearts, and thus do not follow the Gospel wholeheartedly.

In fact, the link between 2 and unreliability is known to most cultures, beginning with the double-faced Janus in ancient Rome. In Persian, *two-faced* means "false" and *two-colored*, "hypocritical," while the Arabs call a hypocrite "fa-

God's hand extends the 2 tablets of the Law to Moses on Mount Sinai. Miniature from the Parma Mahzor, a Jewish prayer book for holidays, written in 1450.

ther of two tongues," or "double tongued," like the German *doppelzüngig*. Things ambivalent and ambiguous belong, as the root *ambi* ("both") shows, to the same sphere of uncertainty (or *twi*light) as the *di*-lemma.

However, to say something positive about 2, one should remember that in medieval Christian exegesis it can point to the two commandments to love God and one's neighbor. Also among the positive aspects are the 2 tablets of the Ten Commandments, which point to the twofold character of human duties toward God and the neighbor. By extension, religious life can be divided into the *vita contemplativa* and the *vita activa*, personified by Rachel and Leah in the Old Testament, and Mary and Martha in the New Testament. In the circles of the Cabalist Isaac Luria, Leah and Rachel were regarded as two

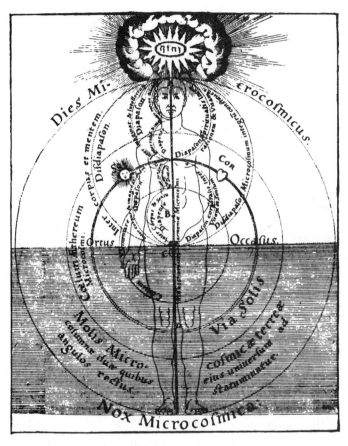

The diurnal and the nocturnal halves of the human being. Diagram from
*Utriusque cosmi, majoris et minoris, metaphysica, physica atque technica
historia* (1617) by Robert Fludd, the spiritual father of Freemasonry. He
claims that God has brought into the world the contrasting aspects of things,
such as light and darkness, form and matter. Hence, according to Fludd, the
world, including man as microcosm, should be explained as a mixture of
opposite principles.

aspects of the Shekhinah, who complains in her exile yet is united with the Lord.

In popular religion, especially in the East, one encounters a widespread fear of the number 2: one should not do 2 things at the same time, marry off 2 couples on the same day, or have 2 brothers marry 2 sisters. Nor should 2 related families live in one room. Jewish law tells man not to pass between 2 women, 2 dogs, or 2 pigs, and 2 men should not allow any of these creatures to pass by them. The Christian peasants of Egypt never have 2 children baptized on the same day in the same church for fear that one will die. Similar superstitions are found in certain parts of the Balkans, where it is held that 2 persons should not drink at the same time from the same fountain.

With this aversion to twofold appearances or actions it is understandable that twins are surrounded with an aura of mystery, and in a number of ancient civilizations twins, or at least one of them, were killed. Among some Northwest Coast Indians, the parents of twins had to observe certain taboos, for it was believed that twins rule over water, rain, and wind and that their wishes are fulfilled by supernatural powers. Among certain Bantu tribes, twins are considered to bring rain.

The special role of 2 is also evident from linguistic forms: besides the use of the dual, groups of 2 are often singled out by special terms, beginning with *twins* (German: *Zwillinge*). Two oxen are called a yoke (German: *ein Joch Ochsen*), and 2 partridges are a brace; one speaks of a *pair* of shoes and a *couple* of things, and of course, the *duo* and the *duet* in music. The German word *Zweifel* "doubt," contains the root "two," as does the Latin *du-bius* and its derivations, including *doubt*. Numerous compounds with the Latin roots *di-* or *dis-*, such as *dispute*, *discord*, or *disagreement*, point to the di-visive quality of the 2.

Although duality and, in a positive sense, polarity are

necessary for the continuation of life, in the worldview of numerologists 2 is more on the negative side, on the side of discord and division (*Zwietracht* and *Entzweiung* in German), and another power is necessary to overcome the disunited world. Thus, one proceeds from the holy simplicity, *Einfalt*, through *Zweifel*, doubt, to the Trinity.

THE EMBRACING
SYNTHESIS

○ ○ ○ ○ ○ ○ ○ **3**

You have 3 guesses: why does the proverb say "Good things come in threes," and why does the psychologist think that 3 restores the damage 2 has caused by dividing? The reason, as Ludwig Paneth has stated, is that the triad leads to a new integration, one that does not negate the duality preceding it but rather, overcomes it, just as the child is a binding element that unites the male and female parents.

Three stands beyond the contrast caused by the 2, as one can still see in the expression of the "laughing third person" and as the German rhyme claims: "Wenn sich ihrer zwei streiten um ein Ei, steckt's der dritte bei." (When two fight for an egg, the third will get it.)

The mysterious character of 3 has often been expressed in poetry, a notable example of which is the sixteenth-century French thinker Du Bartes's *Sepmaine*. In Joshua Sylvester's 1578 English translation, 3 is described as follows:

> The eldest of odds, God's number properly . . .
> Heaven's dearest number, whose inclosed center
> doth equally from both extremes extend,
> the first that hath a beginning, midst, and end.

Indeed, it has each of these aspects and is therefore highly desirable. It can even be seen as the first "real" number, and the first to produce a geometrical figure: since 3 points enclose the triangle, it is the first plane figure that can be perceived by our senses.

Soon after World War II a theologian in Marburg, Wolfgang Philipp, composed a learned study with the title *Trinität ist unser Sein* (Trinity Is Our Being) in which he pointed to the tendency—apparently inherent in everything created—toward grouping things in triads. Indeed, in 1923 Ernst Cassirer had already written in his *Philosophie der symbolischen Formen* ("Philosophy of Symbolic Forms"), that "the problem of Unity which comes out of itself and thus becomes something else and a second to become united with itself again in a third stage belongs to the spiritual heirloom of the entire human race." Although it may appear to be a purely intellectualized expression in the speculative history of religion, the widespread idea of a triune deity shows that it must have deep-rooted foundations in the human mind. As early as the Middle Ages, Albertus Magnus also claimed that 3 is in all things and signifies the trinity of natural phenomena.

For Wolfgang Philipp, all being consists of a tripolar *Ergriffenheit* (emotion), which is manifested in wave, radiation, and condensation, and he thinks that because we are existentially tripolar, we feel at home in corresponding triads. That is why three things are good things, and we are in a good mood when we find our own active, middle and passive principles fulfilled and confirmed in them.

In 1903 the German scholar R. Müller tried to explain the importance of 3 in tales, poetry, and visual art and argued that the importance of the triad stems from the observation of nature. Once human beings saw water, air, and earth, they developed the idea of the existence of 3 worlds (called Midgard, Asgard, and Niflheim in the Germanic tradition); they recognized 3 states (i.e., solid, liquid, and gaseous); they found 3 groups of created things (minerals, plants, and animals) and discovered in the plants, root, shaft, and flower, and in the fruit, husk, flesh, and kernel. The sun was perceived in a different direction and form in the morning, at high noon, and in the evening. In fact, since the world we see and live in is 3-dimensional, all our experiences take place within the coordinates of space (length, height, width) and time (past, present, future). All of life appears under the threefold aspect of beginning, middle, and end, which can be expressed in more abstract terms as becoming, being, and disappearing; a perfect whole can be formed by thesis, antithesis, and synthesis. There are also the 3 primary colors, red, yellow, and blue, from which all other colors can be mixed.

From time immemorial, thinkers have tried to explain the unfolding of the One into multiplicity with special reference to the 3. Lao-tzu says: "The Tao produces unity, unity produces duality, duality produces trinity, and the triad produces all things." The Pythagoreans likewise postulated that the unqualified unity was divided into 2 opposing powers to create the world and then into tri-unity to produce life. For Dante the 3, as he saw it incorporated in the Trinity, revealed the principle of love, that is, the synthetic power. In the history of religions this role of the 3 has led to the formation and invention of numerous trinitarian groupings and tricephalic deities. As early as the third millennium B.C.E. one finds the Sumerian deities Anu, Enlil, and Ea, corresponding to heaven, air, and earth, while ancient Babylon worshiped the astral trinity of Sin (moon), Shamash (sun), and Ishtar (Venus), with 4 planetary deities added to this highest trinity to attain the sacred 7. The Greek goddess Hekate appears under 3 different aspects: in the sky as Selene or Luna, the moon, on earth as Diana, and in the netherworld as Hekate.

In a similar way, hymns to the Virgin Mary in the Christian tradition address her alternately as mother, Virgin, and queen. Although they are not very pronounced in the *Rgveda*, triads of gods are known from ancient India, and there are numerous triadic groups of deities, such as Agni, Soma, and Gandharva. The 3 aspects of the sun are as important as the 3 strides of Vishnu that symbolize them. There are also groups of 3 × 11 = 33 deities, which are expanded in the *Mahabharata* into 33,333. The great triad, however, appears with Hinduism: Brahma the creator, Shiva the destroyer, and Vishnu the sustainer; these were later explained as the 3 aspects of the one unattainable reality. In Mahayana Buddhism, especially in Japan, Amida, Sheishi, and Kwannon represent the heavenly powers. The Etruscans too had a divine triad, but better known is the triad of Greek gods mentioned by Homer: Zeus, Athena, and Apollo. In Homer's work triads and their multiples, enneads, are used in connection with divine things or events. As for ancient Iran, the introduction of the god of Time, Zurvan, and later that of Mithras, into the stark dualism of the ancient Zoroastrian religion shows that even there a third power was needed beyond Ahura Mazda, the principle of light and goodness, and Ahriman, the evil principle of darkness. The *Edda* mentions an ancient Nordic triad in a visionary verse that declares, "Odin gave mind, Hönir soul, Lodur light and color." But even earlier, Germanic religion knew of Odin, Wili, and Weh, who overcame the chaotic matter symbolized in the primordial giant Ymir and created the world from the parts of his body.

The triad of deities is known in ancient Egypt as well. In the state religion professed in Thebes, Amon, Chonsu, and the goddess Mut are the 3 main protagonists, while that of the mystery religion consists of Isis, Osiris, and their son, the savior Horus.

In short, the Christian concept of Trinity is perfectly in

The Trinity: Father, Son, and Holy Spirit, who are persons exclusively "in God." Woodcut from a Book of Hours, Paris, 1524.

tune with general trends in the history of religions. Hopper makes this point in *Medieval Number Symbolism*:

> The paramount doctrinal weakness of Christianity, as the Arian heresy testifies, was the duality of the Godhead. The Son was the first step towards a solution, but the addition of a third person, the Holy Ghost, provided indisputable evidence of Unity . . . . That the Father and Son were One was questionable on numerical as well as philosophical grounds. But Father, Son and Holy Spirit were unquestionably One by the very virtue of being Three!

It is said that St. Ignatius of Loyola used to shed tears whenever he saw something in groups of 3 or something threefold, as this reminded him immediately of the Trinity. This mystery of the Tri-unity has been symbolized by Dante in the structure of the *Divine Comedy*, where the literary form of the 3-part *terzine* serves as the perfect vehicle to express trinitarian ideals. By contrast, it is nearly impossible

Hares, symbols of the tri-unity that is always awake, seeing and hearing everything. Their ears form a triangle. These hares are found on a window of the cathedral of Paderborn. Dürer's painting *The Holy Family with the Three Hares* similarly points to the Trinity: one hare places his paw on the other hare's shoulder, pointing to the third one, who is running off (i.e., God's Son turns toward the world of humanity.)

The Trinity in a quatrefoil. Part of the illustrated title page of the English Lothian Bible, *ca.* 1220. God the Father has Christ's features; God the Son is growing out of his lap; between their two heads the dove, symbol of the Holy Spirit.

to represent the Christian Trinity pictorially. Medieval attempts such as the relief of the Trinity in Plau in Mecklenburg show the difficulties, and one can well understand why Pope Urban VIII regarded a representation in the mosaic floor of the cathedral of Hildesheim (Germany) as heretical: spiritual tri-unity cannot be expressed in corporeal form.

Even more than the supreme deities, the lesser gods commonly appear in groups of 3. The Moiras in Greece and the

Norns in Germanic tradition, for example, are both female powers of Fate who rule becoming, being, and annihilation by spinning and cutting the thread of life.

The number 3 is also important in human relations, as can be seen from the *triumvirate*, the rule of three men, and the old academic dictum *tres faciunt collegium*, stipulating that 3 students must be present in a class and 3 people together can make a decision.

In philosophy and psychology, 3 serves as the number of classification: time, space, and causality belong together. Since Plato, the ideal has been taken to be composed of the good, the true, and the beautiful, while Augustine established the categories of being, recognizing, and willing. The Indian *Chandogya Upanishad* likewise mentions several triadic groups, such as hearing, understanding, and knowledge, and in the later Upanishads, the 3 basic values that express the fullness of the one divine being are *sat*, *chit*, and *ananda* (being, thinking, and bliss).

According to the doctrine of the *Zohar*, the world was created from 3, namely wisdom, reason, and perception, manifested in the fathers Abraham, Isaac, and Jacob. For the Cabalists, the uppermost triad of the ten *sefirot* represents the potencies of perception; the medium triad, the primordial powers of spiritual life; and the lowest triad, the primordial power of vitality. Manichaeism knows 3 ways, and the Temple of the Grail has 3 gates, those of right faith, chastity, and humility.

Spiritual activities are often classified in triads; thus, Hegel writes of the three modes of being as *Ansichsein, Dasein,* and *Fürsichsein*. Spiritual activities are divided into thinking, willing, and feeling, and biological processes in the body also appear in threes. Just as medieval alchemy speaks of the 3 matters that work inside humans, modern chemistry classifies matter as acids, bases, and salts. Physics similarly

The 3 Holy Girls. Mold for gingerbread, Munich. Such images go back to the popular Christian veneration of saintly virgins as well as to Celtic mother goddesses. Thus the Irish cult of the 3 Brigids was so strong that Christianity had to tolerate it and slowly transformed the 3 Brigids into the 1 Saint Brigit, who is to this day the patron saint of Irish women.

relies on the tripartite relations between mass, power, and velocity. The division into spirit, soul, and body has been taken over from Hellenism into both Western and Islamic cultures; drawing on the Quran, Sufism classifies the soul according to 3 degrees: the soul that incites to evil (Sura 12:53), the blaming soul (Sura 75:2), and the soul at peace (Sura 89:27). According to the Samkhya school of Indian philosophical thought, matter has 3 qualities (*triguna*), namely *tamas, rajas,* and *sattva* (the dark, the moved, the being).

One of the most charming traditions about the 3 aspects of life has been retold by Rückert in his poetic version of the story of Al-Farabi (d. 950), the Islamic philosopher and theoretician of music who is said to have played 3 different tunes on his lute: with one tune he made his listeners laugh, with another one he made them weep, and with the third he put all of them to sleep.

Triads have even found their way into the absolute monotheism of Islam. The Shiite form of the profession of faith—"There is no deity save God; Muhammad is the messenger of God; 'Ali is the friend of God"—has led to innumerable citations of the triad God-Muhammad-'Ali in poetry and decorative arts. In some extremist Shia groups even Muhammad, 'Ali, and the Persian Salman al-Farisi are taken together as a kind of triad. In the sect of the Ahl-i ḥaqq, which is predominantly located in Kurdistan, 3 is widely used in cosmogonic myths. In the great tradition of Islam, the quranic division of *islam* (surrender) and *iman* (faith) is enlarged by adding *ihsan* (to do good), and thus the tripartite aspect of religion becomes clear. The legal categories are also threefold: *haram* (prohibited), *halal* (permitted), and *mushabbih* (doubtful). The Sufis, who have divided the way of mortals into *shari'a* (divine law), *tariqa* (mystical path), and *haqiqa* (reality), can be compared to their Christian colleagues who speak of the *via purgativa,* the *via contemplativa,* and the *via illuminativa.* The Sufis also know that the one remembering God (*dhakir*) and the one who is remembered (*madhkur*) finally fall together in the act of remembering (*dhikr*), just as lover and beloved are united in the overarching concept of love. The sayings of the Sufis are commonly divided into 3 parts, just as their hearers are traditionally divided into 3 groups: common people, the elite, and the elite of the elite, each of whom is said to understand the truth differently.

In Buddhism, the threefold classification of words of wisdom and teachings leads to forms like the *trikaya*, or "3 bodies" of the Buddha, and the *Tripitaka*, or "3 Baskets" of the doctrine, as well as the 3 sources of salvation, Buddha, *dharma* (the right path of the law), and *samgha* (community).

Three different worldviews or religions are often grouped into a triad. This is true for China with Confucianism, Taoism, and Buddhism, and for medieval Europe when one spoke of Christians, Jews, and pagans. It is still common to refer to Protestants, Catholics, and Jews, a classification reflected in

"The 3 religions are one." *Left:* The Chinese teachers and founders of religions: Confucius, Shakyamuni Buddha, and Lao-tzu. Stone engraving. *Right:* Christian, Jew, and Muslim united in one confession of faith. From Jacob Emden's *Hashimmush*.

the numerous modern jokes featuring a Protestant minister, a Catholic priest, and a rabbi.

In many traditions 3 was considered to mean "much," that is, beyond duality. Thus Aristotle indicates that 3 is the first number to which the term "all" applies. It is cumulative, and it denotes finality: what has been done thrice becomes law. According to Hartner, in ancient Egypt 3 was, at some point, the upper limit of exact counting and at the same time an expression of indefinite multiplicity; the Egyptian sign for the plural, moreover, is 3 strokes. From this viewpoint, 3 can mean a superlative, as in *terfelix* (thrice happy) or *trismegistos* (thrice greatest one), and the threefold repetition of a word is used for the superlative as well. The trident and triple thunderbolt are attributes of greatness for the deities of antiquity. As a number of perfection and completion, 3 also played a role in the sacrificial rites of ancient Greece and Rome. On special occasions a deity was offered 3 animals, such as a pig, a sheep, and a bullock, or a pig, a buck, and a ram. The same idea occurs in the Old Testament when God asks Abraham for three animals, a cow, a goat, and a ram, each of them 3 years old.

By extension, the comprehensive 3 can turn into a simple round number: Jonah spends 3 days in the belly of the whale; the darkness that came over Egypt lasted 3 days; there are 3 men in the fiery furnace, and St. Paul still feels the results of ecstasy 3 days after his conversion. Likewise, the groups of 3 that are found throughout the Bible and other literary sources, both elite and popular, function as round numbers rather than "mystical" ones, whether these are the 3 sons of Adam or Noah, the 3 best knights, the 3 strongest giants, or the 3 best lovers.

In his book *Agnostos Theos*, Eduard Norden quite correctly points out that "the mystical power of the sacred 3 . . . has extended to the formulaic language of religious

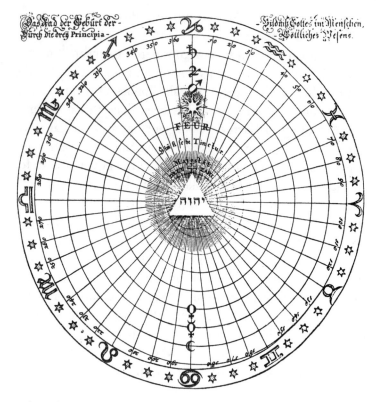

Jacob Boehme, "About the Threefold Life of Man or, Description of the Three Principles of the Divine Being" (written 1618–1619). Large folded folio in the complete edition of *Alle göttliche Schriften* (All divine writings), 1730–1731.

thought." We have already seen some examples from Sufi sayings, but there is also a tripartition of the words of Jesus, especially in the Gospel of John. The Resurrection takes place on the third day, and the resurrected Christ appears thrice to his disciples. It is no accident that there are 3 magi who come to worship the child in Bethlehem, and the epistles of St. Paul and the Revelation are, as Philipp writes, "filled with rolling triads."

Religious buildings are often tripartite as well. This is the case for the Jewish temple, the form of which was later taken over by Christian architects. The long nave, the transept and the apse were then interpreted as pointing to the mystery of the Trinity. The altar too is frequently decorated with a triptych, which usually represents 3 scenes from the life of Jesus. In architectural decoration, especially in Gothic cathedrals, the triskelion is often used.

There were a good number of triads that Christian medieval exegetes interpreted as expressing the central role of the Trinity. Didn't Jesus himself speak of his threefold role as the way, the truth, and the life? And doesn't our very way of conceptualizing and forming sentences prove that the human being is an image of the Trinity, *imago trinitatis*, as Adam Scot thought? Otherwise, how would one think of subject, object, and predicate, or of noun, adjective, and verb, or of past, present, and future when constructing an intelligible sentence? Albertus Magnus saw 3 modes and times of worship and found "that 3 appears in everything and means the trinity of natural phenomena." Our threefold way of acting—by thought, word, and works—is ambivalent: it can lead to positive or negative results, to good works or to sin. For this reason, penance too consists of repentance, confession, and absolution, and the expiation includes prayer, fasting, and almsgiving. The triad of faith, hope, and charity mentioned by St. Paul (1. Cor. 13) manifests itself throughout human life, and indeed, Louis Massignon believed that it distinguishes the 3 great monotheisms, Islam (faith), Judaism (hope), and Christianity (charity).

Medieval exegetes divided history into the time *ante legem* (before the law), which is the time of the patriarchs; *sub lege* (under the law), represented by the period of the Israelite prophets, and *sub gratia* (under grace), the time of the apostles. Thus one could see the progress from the law (Moses) to

Erkennet doch, daß der Herr seine Heiligen wunderlich führet.   Pſalm 4.

Act. 14. v. 17.
Cap. 17. v. 27. 28.

Nachdem alles, was in der H. Schrift aufgezeichnet, uns zur Lehre, Nachforſchung und Erinnerung hinterlaſſen, dabey wir Menſchen unſern groſſen und unbegreiflichen Gott und ſein edles Geſchöpf, alle Creaturen, und zum meiſten uns ſelbſten, recht erkennen ſollen, und dieweil auch in der H. Biebel

Pſ. 104. 148. 150.

der Wunder-Zahlen, Drey, Vier und Sieben öfters Meldung geſchiehet, haben dieſelben auſſer allen Zweifel ein groß Geheimniß in ihnen verborgen.

Darum frage ich in Einfalt, und von reinem Herzen, was doch etliche dieſe nachfolgende bedeuten, ſowohl im Licht der Natur als im Licht der Gnaden.

Von der Zahl Drey.

Erſtlich, was die Drey unterſchiedliche Tage bedeuten, Gen. 8., in welchen Noah aus dem Kaſten den Raben und hernach die Taube alle weg nach verfloſſen dreymal ſieben Tagen ausfliegen laſſen.

Zum andern, was das allererſte Opfer bedeute, ſo Gott der Herr Selber dem Abraham ihm zu leiſten anbefohlen, davon Gen. 15. zu leſen.

Und Gott der Herr ſprach zu mir

Das göttliche Theologiſche Gnaden-Licht,

zeuget vom Natürlichen Philoſophiſchen Licht,

die geiſtliche Sonne

Malach.

die Wurzel Jeſſe,

Bringe mir - und er brachts.
Kuhe
eine dreyjährige Ziegen
—n————n—
Wibber
Cap. 4.

die Wurzel aller Metallen

Und eine Turtel-Taube, und eine junge Taube, · · Adlers Gluten.
und er brachts, und zertheilte es mitten von einander · Solutio Philoſoph.
aber die Vögel zertheilete er nicht, · · · · · · · Sophiſt. Separatio.
und das Gevögel fiel auf das Aas, aber er ſcheuchete ſie davon. Caput Mortuum.

Hermetis Vögelein ☿ friſſet auch die todten Leichnam, und fleucht mit davon, wird endlich vom Philoſopho gefangen, erwürget und getödtet.

Zum dritten, was das für ein heilig wunderlich Feuer geweſen, davon Lev. 9. und 2 Chron. 7. ſo vom Himmel gefallen, die Opfer auf dem Altar angezündet und verzehret. Welch Feuer hernach die Prieſter mit ſich genommen, da ſie in die Babyloniſche Gefängniß weggeführet wurden, und als ihnen der Prophet Nehemias, der auch Jeremias heißt, befohlen, das heilige Feuer in eine Grube zu verſtecken, bis ſie wieder heimkommen würden, hernacher durch die Prieſter wieder ſuchen laſſen; und anſtatt des Feuers, ein dick Waſſer funden, wie aber daſſelbe auf das Holz und Opfer gegoſſen, ſey es von der Sonnen angezündet, und das ganze Opfer zuſamt dem Holze von dem Feuer und Waſſer verbrannt und verzehret worden. Davon 2 Maccab. 1. v. 19. 20. Aus welchem Tages eben dieſes Feuer und Waſſer zu finden, und zu überkommen ſey, welches iſt Prima Materia, in welchen das Gold verzehret wird, und nach der Putrefaction zu einem neuen Leben wieder auferſtehet.

Zum vierten, was die Drey groſſe Wunder-Geburten im Alten und Neuen Teſtament bedeuten, ſo wider und über den Lauf der Natur geſchehen; ſo Gott der Herr ſelber, auch hernach durch ſeine Engel angekündiget und andeuten laſſen. Erſtlich vom dem Iſaac, Gen. 15. 18. & 21. Darnach vom Samſone, Judic. 13. Hernach vom Joh. Baptiſta, Luc. 1. und letzlich die allerwunderſamſte Geburt von unſern Heyland und Erlöſer Chriſto Jeſu, der Jungfrauen Sohn, ſo die andern drey Geburten weit übertrifft, Matth. & Luc. 1.

Zum fünften, was die drey Theile des Menſchen, als 1. der Geiſt, 2. die Seel, 3. der Leib, davon der H. Apoſtel Paulus an 1 Theſſal. 5. ſchreibet, ſowol im Licht der Natur, als im Licht der Gnaden, bedeuten und anzeigen wollen.

Und diß iſt etwas von der Wunderzahl.

The miraculous number 3. From *Geheime Figuren der Rosenkreuzer aus dem 16. und 17. Jahrhundert* (Secret figures of the Rosicrucians from the sixteenth and seventeenth centuries) (Altona, 1785–1788).

the prophets (Elias) and the Gospels. It is remarkable that the founders of each of these 3 traditions were credited with a fast of 40 days. It seems natural that such speculations about triadic processes in history should be developed in later centuries. Joachim of Fiore in the thirteenth century expressed the hope that after the kingdom of the Father and that of the Son, the all-embracing kingdom of the Spirit would begin. This hope was not fulfilled, but the name and the idea of a third kingdom were inherited by the Dritte Reich, the Third Reich, as it was envisaged (though not in the form it later assumed) by several utopian thinkers in early twentieth-century Germany. This *Reich* was, historically speaking, the fourth German Reich, but the symbolic character of the 3 strengthened it and made it more attractive for many. The symbol 3 was stronger than historical truth, for it promised the fulfillment of all human hopes.

The ternary rhythm of historical processes was stressed by the Muslim philosopher Ibn Khaldun in the fourteenth century, and it is also found in church history, where Moscow is called the Third Rome, after Rome and Constantinople. Marx and Comte speak of 3 stages of social development, and it is rather common to find a tripartition of social structures, especially in the Indo-Germanic world, as among early Indians, Iranians, and Celts, where the priest, the warrior, and the agriculturist appear together (although a fourth stratum was later added). Even in modern Germany one used to speak of *Lehrstand, Wehrstand,* and *Nährstand,* the teachers (in the widest sense of the word), the military, and the people who feed others. Expressions like *le tiers-État* (Third Estate) in the French Revolution of 1789 and the modern Third Power and even more, the Third World, belong to the same category.

Another aspect of the 3 is its role as the first geometrical figure, the triangle, which is enclosed by 3 points and formed from 3 lines. As Lüthi remarked: "Plato wanted to build the world from triangles." Although Freud regarded the 3 as the masculine number par excellence (because of the form of the male sexual organ), the triangle or delta was in fact used in primitive Stone Age figures to express the female aspect. The Maya also relate 3 to women, whose characteristics are the 3 stones of the fireplace. Pythagoras interpreted the triangle as the "beginning of the development" in the cosmic sense because geometrical figures such as the rectangle and 6-pointed stars can be formed from it. This important role made the triangle an amulet as well, and people liked to use triangular pieces of paper for magical purposes. Sometimes they had a drawing of the divine eye in the center, sometimes also the Hebrew letters *yod-he-waw* in the points. In the most comprehensive work on German folk religion and superstition, the *Handwörterbuch des deutschen Aberglaubens,* one reads:

> A triangular piece of paper with three crosses in the 3 corners and a prayer in the center helps against gout; triangular pieces of paper at the cradle protect from witches. In 1511, Herzog Maximilian of Bavaria prohibited "blessings written on a certain paper or parchment in triangular shape." In Egypt, children and horses were protected by triangular amulets against the evil eye. In central Europe the linen strainer used to be folded in triangular form to facilitate the churning of butter, and magic triangles were used as well. The socalled triangle of life, triangular numbers, and even triangular cookies also played a role. In the huts of witches all the instruments and implements are triangular.

(However, we would not want to speak of the "eternal triangle" here!)

The double triangle of hermetism, the 6-pointed star, signifies the combination of microcosmos and macrocosmos.

In Christian mysticism and magic, triangles and the number 3 always have a certain relation to the Trinity. But 3 can also become a demonic number, for at times Satan tries to imitate the Trinity and appears in 3 terrible forms. In his *Divine Comedy*, Dante has taken over this motif in a most ingenious way, with Hell becoming a parody, so to speak, of the trinitarian system. (And let us not forget that Peter denied the Lord three times.) In folktales, people often die 3 days after a spirit or a demon has touched them (release from possession by ghosts also takes place on the third day). The intersection of 3 roads (*trivium*) is considered a place of danger, and the gallows tree stands on 3 legs.

In demonic conjurations 3 black animals are sometimes sacrificed (often a buck, a cat, and a dog); in Christian times, demonic animals were imagined to be 3-legged—a demonic transformation of those that had been sacred in the ancient Germanic religion. (The 3-colored cat, however, is regarded as a protective spirit). Odin, who, according to Germanic belief, rides on an 8-legged horse, appears in Christian times on a 3-legged one, and the dogs, badgers, and foxes in his entourage are also 3-legged. The Danish "horse of the dead," and the mounts of spooky demons and of the demon of the plague also have 3 legs, as does the ghost spirit that brings disease to animals. Even the ancient German goddess of death, Hella, rides a 3-legged horse, as a late medieval legend tells. Again according to medieval sources, the devil can appear in the shape of a 3-legged hare. In popular superstition, even birds can have 3 legs, as is often the case in Bavaria and Tyrol for the demonic owl called the *Habergeis*. Should a 3-legged dog run around the house, it means impending disaster. There are even people who avoid sitting at the table in groups of 3, since 3 persons can perform black magic. Thus Shakespeare had 3 witches appear in *Macbeth*, and their incantation is introduced by the pronouncement: "Thrice the brinded cat hath mewed." The witches' question, "When shall we three meet again?" became the key phrase of Theodor Fontane's famous

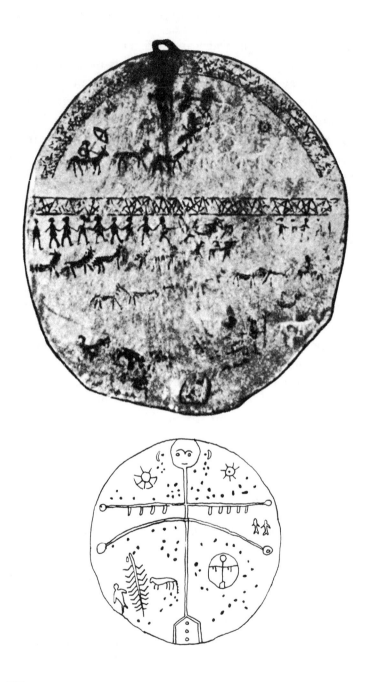

76

ballad, *Die Brück' am Tay* ("The Bridge on the Tay"), with the same uncanny effect: "*Wann treffen wir drei wieder zusamm?*"

Threefold repetitions of certain formulas play a central role in the rites of both religion and magic. In an early formula of the Indian *Rgveda* the enemy is cursed "to lie under all the 3 earths," and the number 3 is found in numerous liturgical acts in both ancient and modern India. The same pattern can be found in classical antiquity, as the poet Horace suggests in an address to the goddess Hekate:

> You that rest near the mountains in forests deep
> and approach the women at their third call in labor
> to save them, virgin, all-powerful,
> three-shaped one!

Even in the *Mirror for Princes* written in Middle Turkic in the early Middle Ages, the *Kutadgu bilig*, the vizier "Highly Praised," goes 3 times to the king "Wholly Awake," and the king is surrounded by 3 counselors.

The 3-fold blowing of the *shofar* in Jewish liturgy is explained by the Cabala as meaning that the first note ascends to heaven, the second one breaks it, and the third one cleaves it.

◀

The 3 worlds of the Eurasian shamans. *Left, above:* Interior of a shaman's drum, Minussian Tatars. Here, the "middle world" is expressed by the netlike horizontal pattern and nine humans immediately beneath it. Nine is the most important power of 3. According to Mircea Eliade, Mongolian mythology speaks of 9 sons of the God, meaning warlike but also protecting spirits. Ulgan, the highest god among the Altay Tatars has 9 daughters, and the skies are often considered to be 9 (See also chapter on 9.) *Left, below:* Exterior of a shaman's drum, Altay people. Vertical in the middle the Cosmic Tree with 9 branches; the upper world is alluded to by the sun and moon, the middle world by 2 human beings, the netherworld by a "helping spirit" (animal), a tree that grows downward, and a smaller repetition of the cosmic tree.

In Goethe's drama, Mephisto reminds Faust, "Du musst es dreimal sagen," meaning that Faust should ask him to enter his room 3 times, otherwise the invitation will not be effective. Blessing and curse are spoken 3 times to be efficient: a haunting literary formulation of this custom is given by Heine in a description of the misery of the Silesian weavers, who weave for their native Germany a shroud filled with a threefold curse:

> O Deutschland, wir weben dein Leichentuch,
> Wir weben hinein den dreifachen Fluch,
> Wir weben, wir weben, wir weben.
>
> (O Germany, we're weaving your shroud,
> We weave into it the three-fold curse,
> We weave and we weave and we weave.)

Similarly, blessings that go back to very early days, such as that of Aaron (Num. 6:24–26), which is now used in church services, often take triadic form:

> The Lord bless thee and keep thee,
> The Lord make his face to shine upon thee,
>   and be gracious unto thee,
> The Lord lift up his countenance upon thee,
>   and give thee peace.

The *trishagion* heard by Isaiah—"Holy, Holy, Holy is the Lord of hosts"—has also become part and parcel of the Christian liturgy, and indeed, Christians make the sign of the cross 3 times.

The threefold liturgical repetition of blessings and prayer formulas is widely known in the Hindu and Islamic traditions as well. At the end of religious ceremonies, Hindus use the threefold *shanti shanti shanti* (peace), a formula taken up by T. S. Eliot in his *Four Quartets*. Muslims make oaths by repeating "By God!" with three different expressions—*wal-*

lahi, billahi, tallahi—and the Quran prescribes a 3-day fast
for expiation (Sura 3:92).

Medieval folklore also utilized the magic number 3 in various
ways because it is the smallest unit from which a magic
square can be made (with 3 numbers on each side). A 3-flower
blessing (commonly calling for roses or lilies) for stanching
blood or catching thieves is first noted in a French document
of 1429 and known in Germany, it seems, from the sixteenth
century onward. One such blessing ("Es standen drei Rosen
auf unseres Herren Gottes Grab . . .") says: "There were 3
roses on our Lord God's grave; the first one is mild, the
second one is good, the third one stanches your blood." There
is also a 3-brothers' blessing, known in writing from the
twelfth century, which usually is part of a story beginning
with the Latin sentence: "Tres boni fratres per unam viam
ambulabant" (Three good brethren were walking together on
one path. . .). The 3 that walk together are usually nameless,
but sometimes they are supposed to be 3 apostles, and in
Byzantine lore they are 3 brothers who have been initiated by
the Holy Spirit into the art of healing. They do this without
taking money and often collect healing plants on the Mount
of Olives and even on formerly sacred mountains like Mount
Olympus. The *Dreikönigssegen,* a blessing connected with the
3 Magi, which was recommended by Pope John XXII in the
thirteenth century, remains in use even today in Germany
and other German-speaking areas. In the same regions one
still sees the letters $C + M + B$, Caspar, Melchior, and Balthasar, written over the doors of houses and stables; this is
usually done by special groups of children or teenagers on
Epiphany, 6 January. The protection of the 3 Magi is also
requested for travelers in a fifteenth-century blessing: "Caspar me ducat, Balthasar me regat, Melchior me salvet at vitam
eternam me perducant." (May Caspar lead me, may Balthasar

guide me, may Melchior protect me, and may they lead me to eternal life.)

Oracles are often based on the use of 3 as well; thus, in divination by arrows in pre-Islamic Arabia, 3 arrows were used. Furthermore, all healing formulas should be repeated thrice: when one has been bewitched—according to an old German saying—one should recite a prayer beginning with the words:

"Drei falsche Zungen haben dich beschlossen . . ."

(3 false tongues have enclosed thee, 3 sacred tongues have spoken on thy behalf.)

And when going to court, one should protect oneself with the formula:

"Ich trete vor des Richters Haus . . ."

(I am going to the judge's house; 3 dead men are looking from its window: one has no tongue, one has no lung, the third one is ill, blind, and mute.)

Numerous traditions in rural areas make use of threefold repetition: in order to accustom an animal to the house, one should lead it thrice around a table leg or make it look in the mirror 3 times, and counting money on 3 holy nights (Christmas Eve, New Year's Eve, and Epiphany) is a way to assure that one will not lack money all year long. In Turkey, a guest is commonly expected to stay for either 3 days, 3 weeks, or 3 months, while the fish, it is said, begins to smell after the third day.

Questions and riddles are posed 3 times or in tripartite form. There is even a game called "3 Questions Behind the Door." The model of the three-part riddle is probably traceable to the riddle of the sphinx: "What goes first on 4 legs, then on 2 legs, and finally on 3 legs?" (Answer: man!) Three is, in any case, an important number in ancient Egyptian

The pillar of the Mauryan emperor Ashoka, ca. 240 B.C., with its 3 lions, model for the escutcheon of the Indian Union since 1949.

tradition, according to which, for example, the shipwrecked man is alone with his heart for 3 days and the serpent god questions him 3 times.

Fairytales often mention 3 human beings, animals, or objects. As a rule, the third and youngest son or daughter is, in the end, the lucky one (as in *Puss'n Boots*). In a German tale the girl asks the 3 animals in the house where she is imprisoned:

> Schön Hühnchen, schön Hähnchen, und du schöne bunte Kuh . . .
>
> (Pretty little chick, pretty little cock, dear pretty checkered cow—what do you think about this?)

Threefold repetitions of places or of stations on the road are also common in fairytales, and very frequently the hero has 3 free wishes: as a rule the third wish is used to obliterate what one has thoughtlessly wished the first 2 times. Similar structures are found in jokes where the third participant outwits the other 2, by either his intelligence or his stupidity. Events last for 3 days and 3 nights, or else for 3 months or 3 years. Sometimes the hero is offered 3 drinks—milk, water, and wine—and is asked which one he prefers. This motif has even entered popular Islamic tales about Muhammad's heavenly journey. The triangular action among hero, enemy, and helper which is rooted in normal life, becomes a favorite motif in fairytales. Threes are also common in folk poetry. Among German folk songs, for example, one can find allusions to the Trinity in "Es blühen drei Rosen" (3 roses grow from one branch . . .), or a girl complaining in distress: "Drei Lilien, drei Lilien, die pflanzt' ich auf mein Grab" (3 lilies, 3 lilies I planted on my grave . . .), or the soldier who sees 3 ravens that bode evil: "Drüben am Wegesrand sitzen drei Raben . . ." (There at the roadside I see 3 ravens—shall I be be the first one buried?).

From the Anglo-Saxon orbit come songs like "There Were

3 Jovial Welshmen," with its repetition, or "The Tourna-
ment," which tells us:

> There hopped Hawkyn,
> There danced Dawkyn,
> There trumped Tomkyn . . .

And nonsense poems take up the ternary rhythm as well:

> Three young rats with black felt hats,
> Three young ducks with white straw flats,
> Three young dogs with curling tails,
> Three young cats with demi-veils
> Went out to walk with two young pigs
> In satin vests and sorrel wigs . . .

Folk poetry, childrens' rhymes, and doggerel often contain
threefold repetitions, of either words, parts of the sentence, or
sometimes an entire line. Formulas like *hop hop hop* or *hip
hip hurrah* or *kling klang kloria* are found everywhere. In
Turkish poetry, a threefold rhyme scheme (abab, cccb, dddb,
etc.) creates a mnemonic pattern, while the repetition of
words clearly emphasizes the meaning, as in the German
nursery rhyme "Ein Schneider fing 'ne Maus":

> A tailor caught a mouse,
> A tailor caught a mouse,
> A tailor caught a mouse,
> A tailor caught a me-mo-mouse . . .

Alan Dundes has compiled a very extensive survey of the
number 3 in American folklore, and his findings can be applied to other areas. Although the use of abbreviations and
acronyms in groups of letters (beginning with the ABC's!)
seems to be more widespread in North America, in the German context one may also think of the abbreviations of party
names, such as CDU, FDP, and FDJ. Dundes's catalogue of the
ternary culture offers the 3 daily meals including a main meal
that usually has 3 courses; the 3 phases of traffic lights, base-

ball's 3 bases, 3 strikes, and 3 outs, the ambitions of sportsmen to perform a triple salto or triple Rittberger, as well as the hat trick—the threefold success in a sports competition, or the 3 grades of Olympic medals.

When American children say in their counting rhymes:

Three little kittens lost their mittens

when French children grow up with verses like:

Quand trois poules vont au champ
(When three hens go to the field)

and German ones with:

Drei Gäns' im Haberstroh
(Three geese in the oat-straw)

the threefold division becomes perfectly natural to them, whether later in their lives they pour out "blood, sweat, and tears," or enjoy "wine, women, and song"; whether the boy remains what is called in German *Dreikäsehoch* (3 cheeses high) or grows to be a stout, "3-bottle man." And at the end, 3 handfuls of dust are cast upon his coffin.

On the literary front, even a quick glance at the titles of books will discover a preference for 3: *Three Men in a Boat; Three Came Back; Three Against the World.* This preference reveals itself not only in the world of belles-lettres but among innumerable scholarly books with titles like *Three Centuries of.* . . . Often collections offer 3 novels, 3 plays, or works by 3 authors. There are more than 1200 titles beginning with "Three" listed in the 1990–1991 edition of *Books in Print*!

Traditional ideas concerning the 3 are also preserved in expressions and technical terms, as Menninger has shown. Thus, the *Drillich* (English ticking) is a coarse material woven with 3 strands of yarn. From the ancient *trivium* (crossroad), but also the 3-part lower education consisting of grammar, dialectics, and rhetoric, comes our word *trivial*. (The *quad-rivium* comprised the higher sciences: arithmetic, geometry, astronomy, and music). Not only the French *travail* (work) but also the English *travel* is derived from the name of an old instrument for torture, the *tripalium*—which may well express some people's feelings.

But let us conclude on a more cheerful note. *Drei-mäderlhaus,* the operetta about Schubert's youthful days with three young girls, brings us to the role of 3 in music. Here we find not only the harmonious triad, but also the ideal tripartite form of sonata and symphony, the string trio, and three-part minuets. In Indian music it is the *tintal,* a rhythm based on the ternary system (although difficult for nonspecialists to analyze) that prevails. But the most delightful unfolding of the ternary rhythm is the waltz, which has become the ideal expression of joyous dance and thus strongly contrasts with the fourfold, down-to-earth rhythm of the march.

THE NUMBER OF MATERIAL ORDER

> Vier widerspenst' ge Tiere ziehn den Weltenwagen—
> Du zügelst sie, sie sind an deinen Zäumen eines.
>
> (Four restive animals draw the world's chariot—
> You bridle them, and they become one with your
> bridling.)

This is how Rückert elaborates on a thought from Rumi, in which the 4 animals of Ezekiel's vision and the Revelation of John are combined with the 4 elements, whose power—manifested in created matter—has to bow down before the one God.

Four is inseparably connected with the first known order in the world, and thus points to the change from nature to civilization by arranging a confusing multiplicity of manifestations into fixed forms. From the evidence of the most ancient scriptures, it is likely that the earliest human beings observed the 4 phases of the moon—crescent, waxing, full, and waning—which thus served as the organizer of time. Likewise, the position of the sun and movement of shadows helped humans to orient themselves in space. By careful observation of the points of sunrise and sunset, especially at the vernal and autumnal equinoxes, the 4 cardinal points were

The Aztec myth of hell according to the Codex Borgia (from the Cholula Tlaxcala region). In the center, a skull surrounded by blood, symbolizing the realm of the dead. At the sides, the deities of the 4 directions, north on the top, south at the bottom; at the corners, punished sinners.

discovered. Together, the 4 directions and the 4 winds provide coordinates for all of earthly life. This was true for the cosmologies not only of Asia and Europe but pre-Columbian America. In Mayan thought, everything is related to the 4 cardinal points, which are, in turn, identified with colors. The cardinal points were represented by a cross whose extremities touched the 4 horizons. Settlements were oriented according to this square: 4 roads led out from the sacred tree in the center along the 4 directions, and at the exit points 4 shrines were dedicated to the guardians of the limits of the village.

In many early cultures the earth was represented by a rectangle. Thus, for the Chinese it took the form of a canopied travel wagon, and this rectangular shape inspired the disposition of fields, houses, and villages according to the principle of *fang*, the square.

In the biblical tradition, the relation of 4 with the 4 directions is expressed by the 4 angels or cherubim, taken to represent God's power extending over the whole world. There are also the 4 creatures of Ezekiel's vision, the man, lion, bullock, and eagle, seen close to the divine throne. In later times these came through a mysterious process to be combined with the 4 letters of the divine name, *YHWH*, the tetragrammaton that must not be pronounced.

The cross, meanwhile, which was obtained by connecting the 4 cardinal points, and was sometimes inscribed in a circle, developed a special religious meaning. In ancient Egypt it became, in a slightly changed form, the hieroglyph for "immortality." In this shape it was even found on early Christian tombstones, and a modern Coptic painter has represented Golgotha with an Egyptian cross as the symbol of Christ's resurrection. The Neoplatonic thinker Porphyry, who found this sign in Egypt and Greece, called it the symbol of the spiritual in the cosmos.

In ancient Europe, the Etruscans arranged their cities and temples according to the cross that gives shape to the world. In their wake, the church fathers saw the cross inscribed in a square as a symbol of the global power of the cross of Christ. Therefore St. Jerome writes: "Ipsa species crucis, quid est nisi forma quadrata mundi" (And the shape of this cross, what is it but the quadrangular form of the world?) The church fathers also believed that the name Adam, if written in Greek letters, alluded to the names of the 4 directions, *anatole, dusis, arkto,* and *mesembria,* and thus Adam became a microcosmic representation of the fourfold material world.

The symbolism of 4 on a breviary of the Zwiefalten monastery, twelfth century. In the center the Lamb of God with the cross, surrounded by the 4 evangelists and their symbols, and the 4 rivers of paradise. In the medallions at the 4 corners, images of the cardinal virtues: wisdom (*sophia*), bravery (*andreia*), temperance (*sophrosyne*), and justice (*dikaiosyne*). These 4 virtues, known from Plato's philosophy, were extended by Thomas Aquinas to 7 with the addition of the 3 "divine virtues."

The relation between the cross of Christ and the 4 corners of the world is expressed in the plan of the ancient Byzantine church, a square with a Greek cross inscribed in it. The 4 arms of the cross (which are of equal length) carry a vault that leads to the central dome. The connection of the 4 with the directions and the created world also finds expression in effigies of tetracephalic deities. In the case of Brahma and other Indian deities, for example, their 4 heads symbolize the 4 directions of the world; Shiva with his 4 arms belongs here too, for in his dance he destroys and recreates the world.

Four has still more aspects, but almost all of them lead back to some relation with the materially established world. However, one should not forget that in many civilizations 4 is an upper limit of counting—one span is 4 fingers, 4 breadths of a palm give one foot. Linguistically, the numerals before 4 are often treated differently from those that follow it, and new groups of numerals or of counting forms begin with 4.

In the classical tradition, a strong interest in the 4 is expressed most clearly in the teachings of the Pythagoreans, for whom it was the ideal number. Its connection with the material world, however, was expressed by the fact that the fourth solid body, the cube, was regarded as belonging to the earth, and thus it was possible to explain the 4-footed beasts as typical earthbound creatures.

The 4 offers a geometrical form that is clear and easily recognizable; therefore tetragonal forms, especially the square, were considered to be perfect and self-contained. In English one still speaks of a "square man," and it may well be that Nietzsche was thinking along similar lines when he expressed his hope for the ideal man who should be "rectangular in body and soul." For the Pythagoreans, however, the *tetraktys*, formed by $1 + 2 + 3 + 4 = 10$, became the all-embracing great unity, and 4 and 10 stand in close relationship to each other.

Amon-Re, the ancient Egyptian "Lord of the 4 directions."

The four-armed dancing god Shiva; at the bottom, the bull Nandi, at the borders two *rishi*s (saints). From a Tamil work about the 64 pleasures of Shiva (*Tiruvilaiyadarpurana* of Paranjoti-munivar, Madras, 1866).

According to Agrippa von Nettesheim (1486–1535), the point where the diagonals of a square meet corresponds to the human navel: *Von dem Verhältnis, dem Maß und der Harmonie des menschlichen Körperbaus* (On the relations, the size, and the harmony of the human body).

Drawing on Pythagorean ideas, both church fathers and medieval European philosophers discovered numerous groups of tetrads, from the 4 elements, which again serve an ordering function, to the 4 temperaments, which explain the variegated world of human psychological forces. In his book on the *tetraktys* and the *decade*, Theon of Smyrna enumerates 10 groups of things that appear in fours: the numbers (units, tens, hundreds, and thousands); the simple bodies; the figures of simple bodies; living beings; communities (individual, vil-

lage, town, nation); capacities; seasons of the year; the ages of
man, and the parts of man (body and 3 spiritual parts). The
exegetes, meanwhile, cite not only the 4 creatures at the divine throne but also the 4 Gospels, not to mention their guiding doctrine of the fourfold meaning of the scripture and the
interpretation of the Bible according to historical, allegorical,
moral, and anagogic principles. Christian exegesis had an ideal symbol for the 4 directions in the form of Christ's cross,
with its 4 arms that indicated, for them, the 4 directions in
which the Gospel should be preached but was also related to
the spatial dimensions of height, length, depth, and breadth.
According to medieval Christian interpretation, the 4 rivers in
paradise flow from Moses through the 4 great prophets of the
Old Testament, and from Jesus through the 4 Gospels. The 4
rivers of paradise are also known in Hinduism; there, the
heavenly cow is said to have produced 4 streams of milk from
her 4 udders. The 4 rivers likewise form an important aspect
of Islamic ideas concerning paradise, and for this reason,
many gardens in Iran and Moghul India were divided by 4
canals into the so-called *charbagh* (4 gardens); this motif was
often applied to mausoleums to suggest an earthly representation of paradisiacal bliss.

The Muslim Brethren of Purity in tenth-century Basra
(Iraq) strongly relied upon Pythagorean ideas and therefore
also stressed the importance of 4. Hadn't God himself formed
the majority of things in nature in tetrads? There are the
qualities of hot and cold, dry and humid; the 4 elements; the
4 humors; the seasons; the directions; the winds; the directions in relation to the constellations; the minerals, plants,
animals, and humans; the 4 types of numbers, and so on. In a
similar vein, a German mystical thinker of the seventeenth
century, Erhard Weigel, found that in the number 4 the *ein-
maleins* (mathematical scheme) created by nature was much
easier to perceive than in the decimal system.

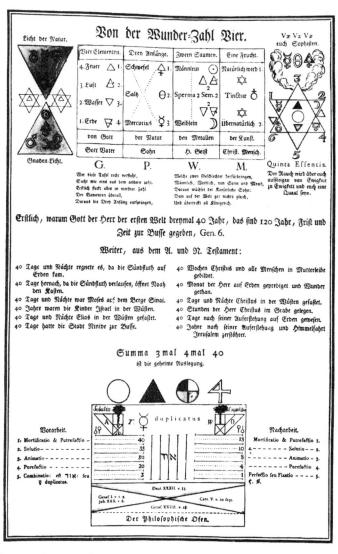

The mirculous number 4. From *Geheime Figuren der Rosenkreuzer* (Altona, 1785–1788).

The patristic thinker Irenaeus went so far as to claim that 4 was such a perfect number of spatial and temporal order that it would be impossible to have more or less than exactly 4 Gospels. Interestingly enough, the 4 sacred writings are not limited to the Christian tradition: in India there are the 4 Vedas, and Islamic tradition designates the Torah, Psalms, Gospel, and Quran as the 4 sacred books.

The 4 plays a prominent role in other areas of Islamic practice as well: a Muslim may have up to 4 legitimate wives; in the case of fornication, the sinner has to make a voluntary confession 4 times (not under duress!), and 4 innocent witnesses must have been present in order to report the act in all details. The first four caliphs from the very beginning of Islamic history are commemorated as the "Righteous Ones" (*rashidun*), and there are, to this day, 4 major legal schools. Even in Sufism, the 4 retains a certain importance: the ascent to the Divine can be divided into 4 steps: the *shariʿa*, or divinely inspired law; the *tariqa*, or narrow mystical path; *haqiqa*, truth, and finally *maʿrifa*, divinely inspired intuitive knowledge. One travels as it were from *nasut* (humanity), through *malakut* (the angelic world) and *jabarut* (the stage of divine power), to *lahut* (divinity). Among the Sufi saints, the 4 saints called *abdal* play an important role, and groups of 4, 40, 400, or 4000 people are often mentioned in legends and tales. And according to Indo-Muslim customs, a boy should first be introduced to the words of the Quran when he is 4 years, 4 months, and 4 days old.

The 4 pervades the Cabalistic vision of the world as well. Indeed, as already mentioned, the universe is divided into 4 parts: *atsilut*, the world of emanation; *beriah*, the world of creation; *yetsirah*, the world of formation, and *asiyah*, the world of visible things. Likewise, the 4 *sefirot* that emanate from the *sefirah binah* play an important role and, according to the Zohar, are related to the 4 directions. Corresponding to the 4 parts of the universe are the 4 colors in medieval tarot, a game that originated in Egypt but found special elaboration in cabalistic circles. A pack of tarot cards still contains 4 colors, black, white, red, and green.

It seems that the concept of 4 as both a numerical unit and the symbol of the world and its order was quite popular in antiquity and the Middle Ages. The 4 goddesses of victory who are found at the base of Phidias's statue of Zeus represent victory over the entire material world.

The 4 is not only a number of spatial order, but one of temporal order as well. Hesiod seems to be among the first known thinkers to speak about the 4 ages of the world (which, after the first, comprising 4800 years, will decrease until the fourth one is only 1200 years long). The Indian tradition knows a similar 4-part development of the world, from the long Golden Age to the last, evil Iron Age, the *kaliyuga*. In Zoroastrianism, one also knew of 4 ages, but their length was equal—3000 years each. On a more individual level, there is the Indian idea of the 4 steps in human life, which culminates in the fourth period by becoming an ascetic—one who does not possess anything. Kabir, the mystical poet of fifteenth-century India who criticized both Hinduism and Islam, complained that there were millions of people seeking the 3 but no one in quest of the 4, meaning that the threefold goal of normal human existence includes *artha* (wealth), *kama* (sensual pleasure), and *dharma* (right behavior), but the only goal worth seeking is *moksha*, freedom from the restlessness of created life. In all these speculations, which seem to be a common Indo-European heritage, one easily recognizes the importance of 3, 4, and 12.

A number of North American Indians, including the Zuni, the Dakota, and the Sioux, have also considered 4 to be a central number in their systems. They know fourfold repetitions, a fourfold division of the years of human life, 4 classes

Four is the perfect number for the Sioux, and the circle is their sacred symbol. Just as there are 4 groups of deities (superiors, allies, subordinates, spirits), there are also 4 species of animals (creeping, flying, quadruped, and two-legged) and 4 ages of the human being (infancy, childhood, maturity, old age). This shield, made about 1866, shows a falcon in the center and 4 feathers representing 4 enemies or 4 major victories.

The importance of the number 4 among the Sioux can hardly be overstated. Their medicine men advised people to undertake all activities in groups of 4. From Royal B. Hassrick, *Das Buch der Sioux* (Cologne, 1982).

of animals, and a classification of heavenly bodies in groups of 4. The number still plays a role in the life of modern Maya in the Yucatán: man is supposed to work in the cornfields, which are arranged according to the cardinal points, and a special ceremony for the integration of a boy into the community takes place in the fourth month of his life.

By virtue of its ordering quality, 4 was always important for practical life, whether among the ancient Kymriens who divided their land into 4 parts or manifold of 4, or in Roman Gallia, where each *civitas* had its *quatroviri*, a governing body of 4 men. Medieval Italian cities similarly had 4 consuls. The Roman arrangement was taken over by Charlemagne and survived in all later court ceremonies, where there were 4 ministries: that of the chamberlain, the marshall or constable, the

Left: Symbolic picture of the Hopi. Wall painting from a *kiva*, a subterranean room for ceremonies, excavated southwest of Albuquerque, New Mexico. In the center a skunk, representing the sun; the outward circle is the sun itself with the 4 fires that emerge from it and burn in the 4 directions in 4 *kivas*. The 4 eagle feathers that emerge from the fireplaces mean "strength." The lines at their ends represent the rays of the sun. *Right:* The symbol of the earth with the 4 directions, according to Sioux mythology. Skan, the Sky, has created the world in groups of four. Sun, moon, earth and sky, i.e., the 4 elements, are round. Day, night, month, and year, the 4-part time, revolve around the sky, and the 4 winds revolve around the borders of the world.

Quetzalcoatl, "Feathered Serpent," culture hero of the Aztecs and best-known Mexican deity, lives through the 4 stages of a corn plant that is exposed to the changing seasons, represented by 4 weather gods. From the Codex Vindobonensis Mexicanus I (*c.* fifteenth century), painted on leather.

seneschal, and the cupbearer (*Mundschenk*). The Ottoman sultan also had 4 main officers in the administration: the vizier, the *qadiasker* (jurist mainly concerned with military cases), the *defterdar* (director of finances), and the *nishanji* (responsible for signing royal documents). In this connection one can think further of the 4 castes in the Hindu system, called *varna*, or "color", and consisting of the brahmins, warriors, settled people, and slaves. In China, there are also 4 categories of people, and the scholar is represented with the emblems of the 4 arts: the guitar for music, the chessboard for logical games, the book for literature, and the picture. The respect for the world-ordering number 4 reached such a de-

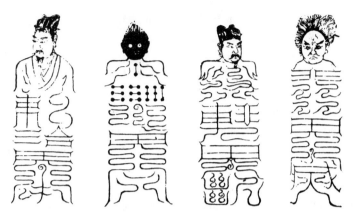

The Masters of the 4 Directions. Chinese amulet to avert curses from ghosts. According to ancient legends, the emperor ordered his supreme fief-holders, called the Four Mountains, to maintain the peace in the 4 directions. China actually had 5 directions; in the fifth one (the center), one has to imagine the emperor's seat.

gree that in medieval Paris and other early universities, professors and students were classified according to 4 "nations"—the French, English, German, and Norman—and students from other nations were assigned to one of these 4 (e.g., the Oriental countries belonged to the French and Holland to the German "nation").

Since the world was imagined to be quadrangular—as John Donne says:

> At the round earth's imagin'd corners, blow . . .

the small picture of the world, the city, has been shaped accordingly. The construction of cities in quadrangular form goes back to the earliest times, as the ruins of Moenjo-Daro in the Indus Valley prove. As early as the third millennium B.C. this city was built in perfect squares. Another, much later example is the city of Hyderabad (Deccan) in southern India, which was founded at the beginning of the second millennium of the Muslim era, in 1591. It was constructed around a center occupied by the Char Minar, an enormous cubic building with 4 high minarets, and extended into the 4 parts of the royal and the business quarters. Rome too was called "Roma quadrata" because of its shape, and thus the square became the symbol of dwelling and of a civic center, even though many "squares" nowadays are anything but quadrangular. Nonetheless, in medieval cities in central and eastern Europe, the town square preserves the old form perfectly. The "quarter" (German *Stadtviertel*) reminds us of the old division of cities into 4 major areas. Other modern terms that go back to a fourfold division include the *Quartier*, the German word for a dwelling place, and the *Quartal*, a German expression for a period of 3 months (i.e., ¼ of the year), but also, the Roman military *quadra* became *squadron*, German *Geschwader*, and the *quaterna* became in French—through Italian *caterno*—*cahier*, "notebook." There is also the *quire*, 4 sheets of paper together, each of which was folded 4 times (thus the printers' sheet with 16 pages).

The number 4 appears now and then in medieval law. According to the *Lex salica*, someone accused of murder had to throw earth from the 4 corners of the house in the 4 directions; if he was unable to pay the penance on manslaughter, he had to go through 4 trials at court and, if nobody bailed him out, was finally executed.

Since 4 is a very "material" and practical number, it rarely occurs in superstitious customs. An exception is the 4-leaf clover, which is supposed to bring good luck, probably because it is so rare. Following an old German custom, on Epiphany, girls would plant one bulb each in the 4 corners of the house; each bulb was given a man's name, and the one that sprouted first was believed to point to her future husband.

In Persian and Turkish 4 occurs also in some popular expressions, but it seems that this is mainly a case of repetition

The 4 Horsemen of the Apocalypse. Miniature painting from the Apocalypse of Arroyo (northern Spain), early thirteenth century. Between the horsemen on their white, red, black, and sallow horses respectively (the sallow one is accompanied by the devil), the 4 evangelists with their 4 symbols: Matthew and the angel, Mark and the lion, Luke and the bull, John and the eagle.

for emphasis: one grasps with 4 hands or looks out with 4 eyes, meaning that one undertakes something with full energy or looks longingly out. "Sizi dört gözle bekledim" ("I waited for you with 4 eyes") means "I have been waiting impatiently for you." In China, however, the similar sound of the words *she* (4) and *shi* (death) provided a reason to avoid the number 4 (as the 13 is often avoided in the West).

The Book of Revelation has described the 4 apocalyptic riders on their colorful horses, who will destroy the created world when they rush through its 4 corners, leaving only the spiritual world unhurt—with the end of matter, there is no longer any need for the ordering power of 4.

There is, however, still another way of looking at the 4. C. G. Jung has emphasized the importance of this number because he felt that its ordering power was needed to counteract the chaotic development in Germany after 1933, to create a new, positive order in the world. In his view, the archetype of quaternity appeared to ban the archetype *Wodan*, which had incorporated demonic restlessness. The *mandala*, the ancient image of spiritual order with its 4 cardinal points, seemed to Jung to be an ideal symbol of the ordering power that even tried to transcend the general trinitarian approach to the world. Therefore a facetious Catholic theologian, Victor White, O.P., remarked in his book *Soul and Psyche* that "Jung had done for the number 4 what Freud did for sex," meaning that he entrusted a basic psychological function to it.

THE NUMBER OF LIFE
AND LOVE

○ ○ ○ ○ ○ ○ ○ 5

Fünf ist
des Menschen Seele.
Wie der Mensch aus Gutem
und Bösem ist gemischt, so ist die Fünfe
die erste Zahl aus Grad' und Ungerade.

(Five is
the human soul.
Just as mankind is comprised of both good and evil,
so the five is the first number made up of even and
odd.)

So Schiller writes in his drama *Piccolomini* (II.1), giving expression to the classical meaning of 5: 5 is connected with the human being. Just as humankind is a mixture of good and evil, so 5 is the first number to be made up of an odd and an even number.

Five has usually been connected with human life and with the 5 senses; it is the number of love and, sometimes, of marriage, as in Plato's *Laws,* where there are 5 guests in the chapter on marriage. The clearest expression of this role can be found in George Chapman's *Hero and Leander* (1598), where an epithalamion, or wedding poem, states:

Since an even number you may disunite
In two parts equall, nought in middle left,
To reunite each part from other reft:
And five they hold in most especiall prise
Since 'tis the first od number that doth rise
From the two formost numbers' unitie
That od and even are; which are two, and three,
For one no number is: but thence doth flow
The powerful race of number. . . .

As an indivisible combination of the masculine 3 and the feminine 2, 5 appears, for pure mathematical reasons, to be a fitting number to express the union of male and female. (The riddle of Turandot in the Persian poet Nizami's epic *Haft paykar* shows that this interpretation was also known in medieval Iran.) Jung regards 5 as the number of natural man: the body and 2 plus 2 extremities make 5 parts. And Goethe, with his usual deep insight into the ancient symbolism of numbers, suggests (in his novel *Die Wahlverwandtschaften*) a temporal marriage for 5 years, "a beautiful odd, sacred number," contrasting with the social family based on the 4 and enclosed in "table, bed, house, and yard."

From early times 5 was considered a somewhat unusual, even rebellious, number, and the discovery by Hippasos of a fifth geometrical body, the pentagondodecahedron, which consists of 12 regular pentagons, embarrassed the Pythagoreans, who had concentrated on the 4. Legend tells that the discoverer of this new body was drowned for his transgression.

It is a remarkable fact that 5 as a structuring number does not seem to appear in crystals (where only 2, 3, 4, and 6 are found) nor did the Pythagoreans use it in connection with musical intervals. In contrast to these absences, however, 5 seems to be the most typical structuring number in living nature, especially in plants. The number of petals is often five, and psychologists such as Paneth have claimed that the opening of buds in the spring, when innumerable 5-starred blossoms suddenly appear, is something "revolutionary." In this connection one has to look at the 1658 poem *The Garden of Cyrus* by the learned English poet Sir Thomas Browne in which "The Quincunciall Lozenge or Network Plantations of the Ancients, are Artificially, Naturally, Mystically considered." Browne tried to show that the number 5 not only pervaded the entire horticulture in antiquity but recurs throughout the life of plants as well as in the figurations of animals, such as the 5 fingers and the 5 toes, or the starfish.

Five could therefore be regarded as a "naughty," unruly number that, like Eros, stirs up the well-ordered cosmos. In terms of their visual geometry, pentagons can never be perfectly aligned together—together, there is always a little empty space left—but they can be successfully used in connection with other geometrical figures. This "incompleteness" was taken up by Christian exegetes such as Augustine, who drew a connection with the Pentateuch, the 5 books of Moses, since these seemed to lack the final perfection that only came into the world with Christ.

One can also look at the 5 in terms of astronomy. From time immemorial 5 has been regarded as the number of the goddess Ishtar and her Roman "successor," Venus. A Swiss scholar, Dr. Martin Knapp, has shown in his *Pentagramma Veneris* how he continued Kepler's classical operations with Saturn and Jupiter in terms of Venus. The upper conjunctions of Venus according to their places in the zodiac in an annual circle correspond to the dates of 2 February 1922, 10 September 1923, 24 April 1925, 21 November 1926, and 1 July 1928. Connecting these dates on the annual circle yields a pentagram, the very sign of Ishtar, Venus, and goddesses related to the planet Venus. The other positions of Venus, such as the lower conjunctions and the eastern and western

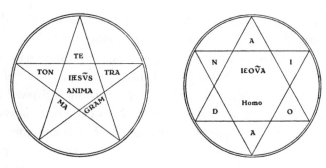

The Pentaculum (*left*) and the Sigillum (*right*) according to Theophrastus Bombastus von Hohenheim, known as Paracelsus. He writes in *De occulta philosophia:* "With these two signs the Israelites and the necromantic Jews have done much and brought about much. They are still kept highly secret by a number of them. For these two have such a strong power that everything that can be done by characters and words is possible for these two." In the magic literature of the sixteenth and seventeenth centuries the concept of the pentacle for all talismanic signs is derived from the pentagram.

elongations, again produce pentagrams when they are connected on the circle, thus confirming the relationship between the 5 and Venus.

The 5 sectors between the points of the pentagram on the circle consist of 72° (72 × 5 = 360), which has given rise to various speculations about the number 72. However, it must be added that the division of the circle into 5 times 72 degrees does not produce the complete synodic orbit of Venus, but rather falls 2.41 degrees short of 8 complete circles. Thus, when exactly computed, the Ishtar pentagram is not completely closed. It is precisely this minute opening that, as Goethe has shown in *Faust*, enables the power of evil, Mephistopheles, to enter Faust's study.

The 5 also plays a role in general astronomical processes. The old solar year, which was based on 60 (5 × (5 + 7)) and could be divided into 5 times 72 days, was later transformed into a correct solar year of 365 days; to accomplish this, it was necessary to add 5 new days, which are called *epagomeneia*.

Drawing on an ancient myth, Plutarch tells how these 5 days came into existence: Helios, the sun god, placed a curse on Rhea because she had betrayed him, and this meant that she would not be able to give birth to the fruit of her illicit love affair under either his rule or that of the moon. But Hermes, intelligent as ever, came to her rescue, and by gambling with the moon god, gained the seventy-second part of a day every day. Thus, by the end of 360 days, he had gained exactly those 5 days that are ruled by neither the sun nor the moon and are therefore exempt from Helios's curse. It seems, incidentally, that in ancient times there was a week of 5 days, which survived in the Roman *lustrum*, a measure of 5 years considered to be one week in the year of the gods.

As for the Ishtar pentagram, it is an infinite figure, just as 5 is a circular number (i.e., its powers always end in 5: 25, 125, etc.). It is therefore not surprising that the pentagram is known in the English tradition as the "lovers' knot," symbolizing endless love. Again we have its relation with Ishtar and the other goddesses of love. Its role on the shield of Sir Gawain may have protective meaning. The sacred 5 of Ishtar and Venus is found throughout antiquity, as in the pentagonal temple of Venus in Baalbek from Hellenistic times. With the advent of Christianity, moreover, Mary frequently assumed the attributes of Ishtar/Venus, including the symbol of the pentagram. Like Ishtar, whose power subdued the moon god, Mary can be represented standing on a crescent moon. It is even possible that the rhythm of feasts honoring the Virgin may have followed the pentagram. In fact, Venus was often regarded as Mary's star, and the beautiful old hymn that begins:

> Ave Maris Stella, Dei mater alma . . .
>
> (Hail, Star of the Sea, nourishing mother of God)

reminds us of these connections.

Is it an accident that in the New Testament there are 5

women named Mary? And since new cults were usually established on places where former religions had celebrated their rites, one finds the transformation of Venus's temples into churches dedicated to the Virgin. A good example is the ancient shrine of Venus Eryeina in Sicily, where Mary appears with 7 veils (perhaps reminiscent of the 7 spheres); these veils are lifted only once a year, on 15 August, the day of Mary's assumption. The number 5 is also a central number in Manicheism, where the 5 sons of the primordial man appear, and the 5 elements of light (ether, wind, water, light, and fire) are confronted with the 5 elements of darkness. The Manicheans knew also of 5 parts of the body, 5 virtues, 5 clerical degrees, and 5 vices. Besides the 5 "helpers of the living God," they spoke of 5 angels collecting the souls.

The number 5 does not belong exclusively to the Near East, however. It is important, and in a certain way even more important than in the Semitic tradition, in China. There, the upper point of the pentagram was inscribed with "earth," followed by "water," "fire," "metal," and "wood," to represent the relations between the 5 elements: earth swallows water, water extinguishes fire, fire melts metal, metal cuts wood, wood ploughs the earth. The most widespread magic square, which is centered on the 5 and became especially important in Islamic lore, originated in ancient China. The Chinese also combined the pentagram with the planets Saturn, Mercury, Mars, Venus, and Jupiter and arranged accordingly the directions, the seasons of the year, sounds, parts of the body, different tastes, animals, and colors. The entire life in this world was based on the 5. One knew 5 sacred mountains, 5 kinds of grain, 5 degrees of nobility, and 5 relationships between people: prince–subject, father–son, man–woman, elder brother–younger brother, friend–enemy. One finds 5 virtues, 5 kinds of fortune, 5 moral qualities, 5 classical books, but also 5 main weapons and 5 punishments. The

Above: The 5 Ancient Ones appear at the birth of Kungfutse (Confucius) above the house of the Kung family. This indicates that the future master stands in friendly relations with the elements that are symbolized by the Ancient Ones: the master of the East (element: wood; color: blue), the Master of the South (fire, red), the Master of the West (metal, white), the Master of the North (water, black), and the Master of the Center (earth, yellow). Old Chinese woodcut. *Below:* 5 bats, 5 swastikas, and in the center, the sign for longevity. The symbol expresses the wish for quintuple happiness according to Chinese belief.

111

old scale in music consisted of 5 notes, corresponding to our octave without the fifth and the seventh.

From this frequent use, it is clear that 5 was an auspicious number in ancient China; thus, on New Year's Day the doors were decorated with strips of cloth bearing inscriptions such as: "May the fivefold luck enter!"

Interestingly, a similarly important role for the 5 can be observed among North American Indians. It was also a key number for the Maya, who considered it to be at the center of the 4 cardinal directions—an idea that later occurs in many

Aztec cross as a symbol of the 5 directions. In the center the deity of fire; above East, left South, beneath West, and right North. Together with the signs for the days and other symbols, it develops into the *tonalpohualli*, the ritual calender consisting of 260 days. From the illustrated Codex Féjer-váry-Mayer, fifteenth century.

European and Asian concepts of the cross with its center. There are Maya deities who are fivefold in form; according to Rätsch, "a deity consists of the 5 colors red, white, black, yellow, and bluish green, and these, in turn, are connected with the cardinal points. These colors, their figures and aspects, taken together, represent the deity. Even today the 5 is found in place names in the Yucatan."

The 5 was also important in ancient India, where the 5 points of the cross (including the central one) were considered

The 5-headed monkey Hanuman, immortal, omnipresent, and full of spiritual power. Five is the number of elements in India as well as in China; they are earth, water, fire, air, and ether. The cosmos and all creatures are composed of these 5. According to the Hindu *shastras*, the purest and most important element is air, belonging to Hanuman.

to be protective and to avert evil. In addition to 5 materials for worship and 5 leaves used for rituals, the Indians knew also of 5 destructive domestic implements, i.e., the grindstone, the hearth, the broom, mortar and pestle, and the water pot. Five elders were in charge of village affairs (*pancayat*), and mystical and magic wisdom was preserved in collections called *panj ratna*, the "5 Jewels," a name that was taken over in Muslim India in the famous mystico-astrological work of Muhammad Ghawth Gwaliori, *Al-jawahir al-khamsa*, in the sixteenth century. The relation of such magical practices to the 5 sacred objects of the Sikh community, however, is not quite clear; a Sikh is distinguished by 5 things, called the 5 *k*, that is, *kes, kach, kara, kripan, kanjha* (long hair, a special kind of trousers, knife, sword, and comb).

There are various opinions as to why the 5 was so important and played a central role in many cultures. The tendency to group things in quinary arrangements is omnipresent. Among such groups, the 5 senses certainly have to do with 5 as the number of natural man, whose life, as Persian poets would say, lasts only 5 days. Yet, 5 also appears as an ordering principle for books and words of wisdom: the 5 commandments for lay people in Buddhism, the 5 main virtues mentioned by Aristotle, and especially the 5 books of Moses, the Pentateuch. In Persian poetry, epics are arranged in groups of 5, and the *Khamsa*, or "quintet," of Nizami (d. 1203) remains the unsurpassable model for hundreds of Persian, Turkish, and Urdu poems in this style. The Indian collection of fables, the *Pançatantra* (5 Teachings), follows a similar quinary principle, as does the tendency to connect counsels or wise words to sequences of 5. Similarly, Goethe, in his *West-Östlicher Divan*, offers 2 sets of such advice consisting of 5 sentences each.

But what about the pentagram and the 5 as a protective number?

Hand with 5 fingers, representing the *panjtan* (Muhammad, ʿAli, Fatima, Hasan, and Husayn), with the invocation "Yā ʿAli." Silver amulet against the evil eye. Pakistan, twentieth century.

The human hand with its 5 fingers is especially common as an amulet in the Islamic world. Among Muslims it is called the "Hand of Fatima" after the youngest daughter of the prophet Muhammad and mother of the Imams. At the same time it can symbolize the *panjtan*, "5 people," that is, Muhammad, Fatima, her husband ʿAli, and their 2 sons, Hasan and Husayn, who are the most important figures of Shiite Islam but who are also highly esteemed among the Sunnites. The Hand of Fatima is supposed to avert the evil eye, and the great stone hand over the entrance to the Hall of Justice in the Alhambra is just such a talismanic sign. In popular traditions throughout the Middle East and North Africa, the sign of the hand is woven or embroidered in flags and banners, painted over doorways, and suspended from car mirrors in order to avert evil. Its shape is also reminiscent of the invocation *Yā Allāh*, "Oh God!" in Arabic letters, and hence is believed to provide help. Conversely, when one greets someone by extending the hand, with palm pointed to his face, and saying *khams fi ʿaynak,"* 5 in your eyes," the real or imagined evil eye of the other person cannot work. The very gesture of stretching one's hand against someone's face, or the simple word *khams*, "five," can be understood as a very dangerous curse. For this reason, in spoken Arabic the number 5 is sometimes expressed with a different word or special forms are used so that no curse is effected.

Even more widespread than this sign is the pentagram. It was used as an amulet in the ancient Near East, where the Divine Mother, Ishtar, was believed to protect humans from the evil spirits. The sign is found now and then on the walls of medieval churches, where it served to protect the house of God against Satan, or to ward off the demons who were once worshiped, as positive powers, in the company of the Germanic goddess Perchta.

In practical magic the pentagram is regarded as the sign of the microcosmos, as Paracelsus pointed out. The leading French magician of the nineteenth century, Abbé Constant, who called himself Eliphas Levy, has dealt with the pentagram in detail. For his private use he had—as he called it—a "scientifically perfected pentagram," which was framed not by simple lines but rather by bands of astrological symbols and cabalistic words. On the 5 outer edges was inscribed the word *tetragrammaton*, "with four letters," alluding to the transcription of the ineffable Divine name in Hebrew: *YHWH*. In Constant's view, the pentagram expressed the rule of the spirit over the elements, and it could be used to fetter demons of the air as well as fire and water spirits and the ghostly creatures connected with the earth.

The use of the pentagram is frequent in modern cabalistic and magic practices, especially as the symbol of the microcosmos. Agrippa von Nettesheim, the sixteenth-century Ger-

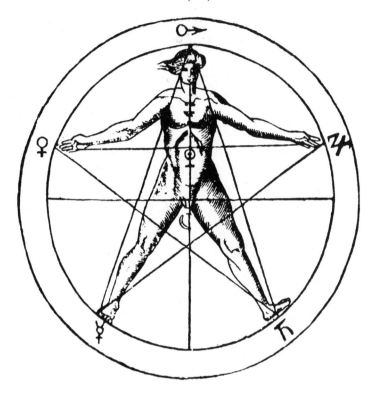

The Pentagram from Agrippa von Nettesheim, *Liber quartus de occulta philosophia* (1535).

man physician, philosopher, and occultist, has shown that when a person stands upright with legs opened and arms extended and slightly lowered, his extremities will cut a circumscribed circle rather exactly on the 5 points of a pentagram. Drawings of this figure, often filled with symbols of the planets and other important signs, occur frequently in later medieval and Renaissance lore.

Since the human being consists of 4 elements, a fifth, secret one (*quinta essentia*) was added in order to reach the sacred 5. This *quinta essentia*, our *quintessence*, was considered to be the real element of life, and its production was a goal of medieval alchemists. (Here one may see connections between alchemical processes and Manichaean cosmology, as Gernot Windfuhr has recently pointed out.) To find the principle of life and to overcome death one has to rely on procreation and Eros, so the *quinta essentia* again points back to the ancient life-giving power of the Mother Goddess through which life is continuously renewed. It seems that the nineteenth-century German poet Eduard Mörike knew, or sensed, these connections, for in his story of the *Stuttgarter Hutzelmännchen* (The little gingerbread man of Stuttgart) the "beautiful Lau" has to laugh 5 times in order to conceive a living child. And then there is the Indian god of love, Kama, with his 5 arrows made of flowers, as is clear from religious dramas such as the *Gitagovinda*.

One cannot say for sure whether the sad Baltic song "Zogen einst fünf wilde Schwäne" (Once there flew 5 wild swans) has something to do with erotic symbolism, but since it tells of 5 youthful heroes who never returned and 5 virgins who waited for them in vain, this interpretation is quite likely.

Christian theologians in early centuries of our era interpreted the 5 in a different way. They saw in it not only the law of the Pentateuch but also the 5 wounds of Christ, and they thought the number was combined from 3 + 2, that is, faith in the Trinity and the twofold command to love God and one's neighbor. In *Against the Heretics*, Irenaeus went so far as to accuse the gnostics of paying inadequate attention to the number 5, in view of the fact that the words *soter* (savior) and *pater* (father) consist of 5 letters each and although the Lord fed 5000 people with 5 loaves of bread; although the cross has 5 main points, and although the 5 fingers and the 5 senses testify to God's acts. It seems, however, that the relation between the 5 wise and 5 foolish virgins and the number of Ishtar was not discovered by the theologians.

Another religious tradition in which 5 is very important is Islam. In addition to the 5 Pillars of Faith (i.e., profession of faith, ritual prayer, fasting during the month of Ramadan, almsgiving, and the pilgrimage to Mecca), the ritual prayer is performed 5 times a day, and the categories in Islamic law are 5: duty, recommended, indifferent, disapproved, and prohibited. In wartime the booty is divided into 5 parts. We have mentioned already the *panjtan*, the 5 members of Muhammad's household who were taken under his cloak and whose names are used in amulets; in addition there are the *panj piriya*, 5 mysterious saintly people who are invoked in Indian Islam. The Shiite brotherhood of the Brethren of Purity in tenth-century Basra explicitly stated that Islam is based on 5—not only because of the 5 Pillars and the *panjtan*, but also because there exist only 5 law-giving prophets (Noah, Abraham, Moses, Jesus, and Muhammad), and among the mysterious isolated letters that are found at the beginning of several chapters of the Quran no group exceeds 5 letters. The traditional ideas about the 5 planets, the 5 additional days, and the 5 geometrical figures were also used by these philosophers as proof of the importance of the number 5. Before the Brethren, or perhaps at about the same time, the Arab philosopher and alchemist Jabir ibn Hayyan cited the Greek philosophical pentads and the 5 platonic bodies as well as the doctrine of Empedocles, who speaks of *materia prima*, intelligence, soul, nature, corporeal matter. Jabir himself divides the created world into substance, matter, form, time, and space. Even earlier, another Muslim philosopher, al-Kindi, the "philosopher of the Arabs" (d. after 870), had postulated a pentad consisting of matter, form, movement, space, and time.

The 5 also plays an unusually important role in the proto-

Ismaili philosophical text the *Umm al-kitab* (Mother of the Book), possibly influenced by gnostic and Manichaean speculations about quinary groups of spiritual beings. The *Umm al-kitab* begins with the 5 lights, each of which has 5 colors; each color is connected with one of the *panjtan*. There are 5 archangels, Michael, Seraphiel, Gabriel, Azrael, and Suriyel, who correspond to the 5 spiritual capacities. (The complicated early and proto-Ismaili gnosis has been studied in detail by Henry Corbin.)

The ancient mystical-magic meaning of 5 (about which Bachofen wrote quite extensively in his work on matriarchal society) and the amuletic function of the pentagram is still alive in popular customs. If 5 bread crumbs are cast on the table, for example, and a cross can be formed by moving just one crumb, then it is believed that a question raised while casting the crumbs will be answered affirmatively. A blessed bouquet of 5 herbs enables its bearer to recognize witches and magicians. According to medieval German folklore, the pentasyllos root can make a person resistant to poison and drives away evil spirits; it also reconciles people. In the Vogtland in eastern Thuringia it was believed that a toothache sufferer should spit 5 times into a bush of yellow willows and then make 5 knots in one of the twigs of this very bush. Once the twig is dried up, the toothache is also over. This is a typical example of the connection between sympathetic medicine and magic practices. In another part of Germany, Hassia, the harvest was begun with a child under the age of 5 cutting the first ears.

The irritating, magical, natural, and erotic qualities of 5 along with the alternately dangerous and protecting power of the feminine have been captured well by John Donne in his poem "The Primrose":

> Live, Primrose, then, and thrive
> With thy true number five,

And women, whom this flower doth represent
With this mysterious number be content.
Ten is the farthest number; if halfe ten
 belonge unto each woman, then
 Each woman may take halfe us men,
Or if this will not serve her turne, since all
Numbers are odd, or even, and they fall
First into this, five, women may take us all.

Attesting to this role of women are a whole range of phenomena, from the 5 wise and 5 silly virgins of the Gospels to
the successful perfume Chanel No. 5.

THE PERFECT
NUMBER OF THE
CREATED WORLD

○ ○ ○ ○ ○ ○ ○ **6**

According to ancient and Neoplatonic systems, 6 is the most perfect number as it is both the sum and the product of its parts: it is formed either by adding $1 + 2 + 3$ or by multiplying $1 \times 2 \times 3$. Furthermore it is the product of the first male (2) and first female (3) numbers. From the psychologist's viewpoint, it represents a combination of analysis and synthesis in its simplest form: 2×3. It summarizes all the plane figures of geometry (point, line, and triangle), and since the cube is composed of 6 squares, it is the ideal form for any closed construction.

Biblical exegetes were delighted to discover these mathematical qualities, for they knew that God had completed his work of creation in 6 days. For Philo and his successors this was by no means surprising, for, as Hrabanus Maurus says: "Six is not perfect because God has created the world in 6 days; rather, God has perfected the world in 6 days because the number was perfect." Augustine argues in a similar vein and is even able to divide these 6 days into 3 parts: on the first day God created light; on the second and third days, heaven and earth, the *fabrica mundi;* and in the 3 last days he made the individual creatures, from fish to man and woman. Fascinated by the meaning of the 6, some exegetes went so far as to read the opening of the Hebrew book of Genesis, *b'reshit* ("in the beginning"), differently in order to obtain the phrase *bara shīth,* "He created the Six."

The doctrine of creation in 6 days led to the arrangement of the week with its 6 working days and 1 day of rest. Logically one also thought of 6 years of work to be followed by a festive year: during the seventh year the land should lay fallow, "a sabbath of rest to the land" so that the poor could eat what grows on the fields without sowing (Ex. 5:4). The fact that the seraphim of Isaiah's vision (Is. 6:2) had 6 wings points to their perfection.

The 6 days of the Creation. Title page of a psalter from Canterbury, early thirteenth century. On the first day the Lord carries compasses and scales (Wisdom of Solomon); in the following days he is transformed into a scholar, lifting the Scripture with great authority; only when creating the plants does he look more relaxed.

Christian tradition not only acknowledges the perfection of the 6 but also connects it with Jesus's crucifixion, which took place on the sixth day of the week and was completed during the sixth hour of the day. This sixth hour (*sexta*) developed later into the Italian term *siesta*, meaning "rest in the afternoon." In Matthew (25:34–36), the 6 is seen as a symbol of the *vita activa*, the life of good works. And following the Old Testament tradition, 6 is seen as a preparation for either rest or completion, which will happen on the seventh day, or year. Thus—as Revelation tells us—6 angels blow the trumpet as long as the Last Judgment continues, and the seventh angel will begin to blow when the divine Mystery is finally completed (Rev. 10:7).

The 6 also occupies a prominent place in Zoroastrianism, where 6 periods of creation are related to the 6 supreme angelic beings, the Amesha Spentas. Again, this spiritual sextet is completed by Ahura Mazda, the god of good, who is the seventh and, at the same time, all-embracing supreme spiritual being. And since there are 6 periods of creation, Zoroastrianism has also 6 great feasts. The connection of 6 with creation is known to the Islamic tradition as well, but there, it often has a somewhat more negative aspect. Mystics and poets experience this world as a cubic cage in which they are imprisoned, struggling in vain to escape the bondage of the five senses and the four elements. Persian poets refer to this situation as *shashdara*, "6-door," which is the hopeless position of the gambler in a trictrac game.

A more positive evaluation of the 6, again connected with the world of creation, can be observed in the Hermetic tradition. Here, the hexagram, which consists of 2 combined triangles, one pointing upward, the other one downward, represents the macrocosm. In the ancient Indian tradition, for example, this figure expresses the union of the creative Vishnu triangle with the destructive Shiva triangle and thus refers

Upper left: The 6-pointed star in the center of the Indian *Varana yantra* contains the *Bija mantras,* which deal with the spiritual personality of the monkey Hanuman, the hero of the Ramayana. The 8 lotus petals around the circle bear the names of the most famous monkey warriors, beginning with the king Sugriva. *Lower left.* Six-pointed vault closure in the church of St. Mary at the Cistercian monastery of Doberan near Rostock, Germany, late thirteenth century. The number 6 also determines Mary's starry robe and her crown. *Right.* Six-winged angel as symbol of anthropos, "man," standing on 2 wheels, the Old and New Testaments. After a painting in the monastery of Watopaedi, Mount Athos, dated 1213.

to the creation and destruction of the material world. The upward pointing triangle was taken as the symbol of positive, good aspects of life while the downward pointing one refers to the material, evil, destructive aspects of the world. The hexagram could be interpreted in terms of polarities between spirit and matter, God and chaos, transcendent and immanent, which again represents the created world and even more, the male (upward) and female (downward) aspects of life. It is not certain to what extent the hexagram was known and used in antiquity; it certainly appears in the Near East, but the association of the "star of David" with Judaism goes back only to the seventeenth century. In Christian mysticism the hexagon similarly symbolizes polarity, but it is also taken to represent half the zodiac. In the portal of the cathedral in Freiburg, Germany, one sees a double row of 6 angels each, which are interpreted as allegorical manifestations of the 6 virtues of Christ. These virtues, if imitated by mortals, form the hexagonal shield of faith.

The hexagon's role in building the world, as it was understood symbolically in ancient traditions, has become much clearer thanks to scientific observation. Indeed, it appears as an ideal building principle in nature, the most obvious examples being the beehive and the snowflake, whose forms have fascinated observers from earliest times. More strikingly, it appears in the molecular structure of the benzine ring, C_6H_6, which the chemist Kekulé discovered in a dream.

THE PILLARS OF
WISDOM

○ ○ ○ ○ ○ ○ ○ 7

Even though one may not have been born, like the author, on the seventh day of the month at 7 A.M. in a house number 7, one could certainly become aware of the importance of the 7 by crossing the 7 seas in the modern version of the *Sieben-meilenstiefel* (the 7-league boot), namely the Boeing 747, and arriving in the area of the Siebengebirge (7 Mountains, near Bonn), or by drinking 7-Up in September, the ancient seventh month.

The number 7 has fascinated humankind since time immemorial. Thus, in a study called *Seven, the Number of Creation*, Desmond Varley has tried, as did many others before him, to reduce everything in the sublunar world to the number 7. Indeed, it consists of a ternary of creative principles (active intellect, passive subconscious, and the ordering power of cooperation) together with a quaternity of matter encompassing the 4 elements and the corresponding sensual powers (air = intelligence, fire = will, water = emotion, earth = morals). Such a division of the 7 into these two constituent principles, the spiritual 3 and the material 4, was used time and again in medieval hermeneutics and is also at the basis of the division of the 7 liberal arts into the *trivium* and the *quadrivium*.

Another book, written at the end of the last century, points out that periodicity is always connected with 7, be it in music, where the 7 notes of the scale return to the first one in the octave, or in the order of chemical elements. To be sure, the author takes up the ancient idea that the growth and development of humans takes place in periods of 7 and 9. The 7 ages of man, cited by Shakespeare, were known in antiquity, and in a pseudo-Hippocratic book 7 is called the number of cosmic structure: the author claims that it causes winds, seasons, the ages of man, and the natural division of the human life. But even before the unknown author of this book, Solon used the stellar spheres to divide human life into 7 periods of 10 steps each, and the ideas of this Greek sage were combined in turn with concepts from the Old Testament by the great Hellenistic Jewish philosopher Philo of Alexandria. Philo writes that at the end of the first heptad real teeth appear instead of the milk teeth; at the end of the second heptad, puberty begins, and in the third, youth sprouts a beard. The fourth heptad is the high point of life and the fifth, the time for marriage. The sixth heptad brings intellectual maturity, while the seventh ennobles the soul through reason, the eighth perfects both intelligence and reason, and in the ninth, the passions are tamed, making way for justice and temperance. As for the tenth heptad, it is best for death, since in its wake people are only decrepit, useless, senile beings: as the Bible attests in Psalm 90, the human life is threescore and 10.

Such ideas were widespread all over the western world, and in the seventeenth century Sir Thomas Browne wrote that every seventh year brings some change in life, in the temper of either the body or the mind, and sometimes even in both. However, 3 points are of special importance, that is $7 \times 7 = 49$, $9 \times 9 = 81$, and $7 \times 9 = 63$. The sixty-third year is the "grand Climacteric . . . which is conceived to carry with it the most considerable fatality."

The idea of human development in stages of 7 years is clearly expressed in the famous philosophical novel *Ḥayy ibn Yaqẓán*, the *philosophus autodidactus* composed by the Arab physician and philosopher Ibn Tufayl in the twelfth century and translated into English as early as the seventeenth century: the hero, growing up completely alone in the wilderness, develops in steps of 7 years to moral and spiritual perfection.

In China, 7 is also connected with human life, and especially with female life: the girl gets her milk teeth at 7 months and loses them at the age of 7; in 2×7 years the "road of *yin*" opens as she reaches puberty, and at $7 \times 7 = 49$ menopause sets in. From a medical viewpoint, this is quite correct, and one can add that menstruation regularly comes every 7×4 days. Likewise, pregnancy is counted by summing up 7×40 days from the first day of the last menstruation. Before modern medicine, it was said that a baby could survive when born in the seventh month of pregnancy but not in the eighth month.

Thus, as the Pseudo-Hippocrates asserts (quoted by Varley): "The number 7 because of its occult virtues, tends to bring all things into being. It is the dispenser of life and is the source of all change, for the moon itself changes its phases every 7 days. This number influences all sublunar things." With this remark Pseudo-Hippocrates indeed touches the possible source for the veneration of 7: the 4 phases of the moon, whose shape clearly changes every seventh day. As the moon god was the highest deity in the pantheon of the ancient Near East, everything connected with him was venerable. Seven appears as the number of planets in Babylon: Sun, Moon, Mercury, Mars, Venus, Jupiter, and Saturn (among which neither sun nor moon are real planets), although this heptad may have been formed in order to reach the sacred number, "and when the seven planets had been found one needed not look for more," as Hopper says. These 7 planets are in turn part of the 7 heavenly spheres, and this idea has stimulated human imagination through the millennia. Thus Goethe sings in his *West-Östlicher Divan*:

> Die Planeten haben alle sieben
> die metallnen Tore aufgetan.
>
> (All the seven planets have
> opened their metal gates.)

The ancient Babylonian *ziggurat*, the step pyramid, had 7 storeys, and the temple of the Sumerian king Gudea was called "the house of the 7 parts of the world," which likewise had 7 steps, to remind the visitor of the 7 spheres. The Tree of Life, too, was represented with 7 branches, each branch having 7 leaves, and it may well be that this was the model for the 7-branched candelabrum of Jewish ritual.

Nor was the 7 sacred only in the Near East: in pre-Columbian America, the Maya believed in a 7-layered sky and considered 7 the number of orientation in space. According to their belief, the conjunction of woman (3) and man (4) produces a unit, 7, which is endowed with life. Although these associations of 3 and 4 are the opposite of those prevailing in Asia and Europe, where 3 is masculine and 4 feminine, they still result in the same truth: 7, the living organism.

The Babylonian calendar, introduced at the time of Hammurabi in the eighteenth century B.C.E., relies upon the phases of the moon, and the seventh, fourteenth, twenty-first, and twenty-eighth days of each month are regarded as unlucky periods during which one should avoid certain activities. In Judaism, the seventh day became the day of divine rest, thus sacred: the previous, negative injunction against work was transformed into a positive order. Such a transformation is often encountered in the history of religions, where sacred and profane, positive and negative orders exchange their position with the introduction of a new religious system. Seven, with its ancient "negative" connotations, is thus to a certain extent ambivalent. Further, it should be remembered that in the hot season in ancient Babylonia 7 signs of the zodiac were invisible and only 5 could be observed above the horizon. It was believed that the invisible 7 signs had gone into the realms of the evil principle, and for this reason, 7 could be considered a dangerous or evil number. In Germany, an extremely nasty and troublesome woman is called *eine böse Sieben*, "an evil Seven," a term that first appears in German sources in 1662. Through this association with evil, 7 was frequently used in magic practices—weren't the witches of Salem, Massachusetts, connected with the House of 7 Gables? And then there is the "darned seventh year" (*das verflixte siebente Jahr*), or in married life, the "7-year itch."

The number of the 7 planets turns up in ever-new disguises in the mythological figures of gods, heroes (*Seven Against Thebes*), or wise men to symbolize heroism and creative wisdom. In addition to the 7 planets, there were also the 7 stars of the Pleiades, which could be seen by the naked eye. These stars were also hidden beneath the horizon in Babylonia during the 40 hottest days and the days of heavy rainstorms. Therefore they too were often considered to be evil-intentioned spirits. From this apparently comes the conception of sevenfold devils, as in Luke 8:2, where 7 devils are exorcized from Mary Magdalen.

Despite its negative possibilities, 7 is usually endowed with positive powers. In Babylonia, its sign designated wholeness and plenitude, and the concept apparently spread out from Mesopotamia to the neighboring civilizations. In Egypt, for example, there were 7 paths to heaven and 7 heavenly cows, and by doubling this number one reached the 14 places in the realm of the dead. In a late demotic text, Osiris leads his father through the 7 halls of the netherworld.

The Old Testament is replete with heptads. In the seventh generation after Adam there appears Lamech, who lives for 777 years and should be avenged seventy-sevenfold (Gen. 4:24). The 7 steps leading to Solomon's Temple correspond to the 7 storeys of the Babylonian temples. Noah's dove stays away for 7 days, and the flood prepares its arrival for 7 days; the Euphrates is divided into 7 brooks. Recompenses and punishments are repeated 7 times, and 7 blessings are part of the marriage ceremony. During the sacrificial expiation in ancient Israel, blood was sprinkled 7 times, and as most great feasts lasted for 7 days a 7-day sacrifice was celebrated when Solomon's temple was inaugurated.

Weddings, like that of Samson with the 7 locks, also lasted for 7 days, as did periods of mourning. (Incidentally, the idea that the soul needs 7 units of time to free itself from the dead body can also be found in other parts of the world, as in China, where rituals for the dead are performed 7 times on every seventh day.) Besides the sevenfold sacrifice, the Old Testament mentions the sevenfold washing in the Jordan (2 Kings 5:10, 5:14) as well as the sevenfold sneezing of a person revived from death (2 Kings 4:35). The ambivalence of the 7 is well expressed in Pharaoh's dream of the 7 fat and 7 lean cows (Gen. 41:1–4) and is also visible in Jacob's serving first for Leah for 7 years, and then another 7 years for Rachel (Gen. 29:18–30).

As 7 contains everything, the Proverbs praise the 7 Pillars of Wisdom (9:1), and when Zechariah speaks of the 7 eyes of the Lord, he uses this image to evoke God's omnipresence and omniscience (4:10). The idea of 7 divine eyes reoccurs in later Sufism in connection with the 7 great saints who, as it were, are the eyes through which God looks at the world.

Seven is then extended into 70: the Old Testament speaks, for example, of 70 nations and 70 judges in the Sanhedrin. Further expanded into the thousands, it yields the 70,000 veils of light and darkness that, according to Sufi ideas, separate man from God, or the idea that God is praised by beings with 70,000 heads, each of which has 70,000 faces, each face 70,000 mouths, each mouth 70,000 tongues, and each tongue 70,000 languages. Similar expressions of marvels can be found in the description of the heavenly virgins, the houris who await the faithful Muslim in paradise.

"The words of Yahweh are pure words, purified silver, purified 7 times," says the author of the twelfth psalm, and these words inspired the Cabalists to a deeper interpretation of biblical expressions. Thus the 7 days prescribed for dwelling in the *sukkot* relate to the 7 days of creation, and according to the *Zohar*, the 7 lower *sefirot* in the Cabala are connected with their historical manifestations in the form of Abraham, Isaac, Jacob, Moses, Aaron, Joseph, and David. Likewise the building of Solomon's temple, which lasted for 7 years, is combined with the 7 lower *sefirot*, which are contained in the *sefirah binah* ("intelligence"), corresponding to the Temple. The seventh of these lower *sefirot* (the tenth one altogether) is the Shekhinah, which is called the Sabbath Queen and which, the *Zohar* explains, corresponds to the seventh primordial day.

The Old Testament (Gen. 4:15) says Cain's murder will be avenged 7 times, but the New Testament (Matt. 18:22) speaks of 70 times sevenfold forgiveness. Rupert of Deutz, a German theologian of the Middle Ages, speculated about this mystery and reached the conclusion that 7 is an indivisible, that is, unchangeable number and thus points to the unchangeable vengeance of the Old Testament as well as to Christ's immutable life, *immutabilis vita Christi.* Conversely, 77 in Gen. 4:24 (revenge for Lamech) is a divisible number, which means that such a vengeance was abrogated by 70 times sevenfold forgiveness. Among other Christian elaborations on the heptads of the Old Testament are the 7 last words of Christ on the Cross and the 7 stars he holds in his hand. In the Revelation of

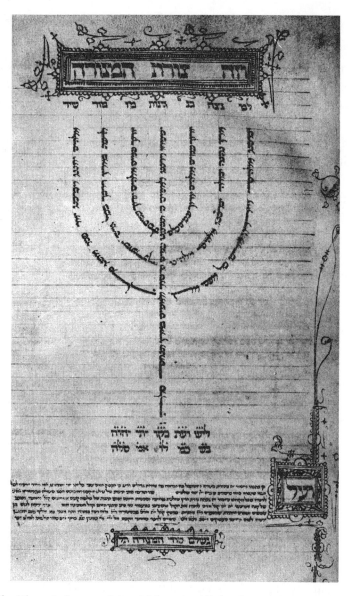

The 7-branched *menorah* (candelabrum) is filled with symbolic meaning. In this picture, the words of Psalm 67 are used to form the *menorah*. As each Hebrew letter has a numerical value (1 to 22), the letters of the 7 branches of the candelabrum can be added and then retranslated into words. The results of this esoteric game are explained in the inscription: the final result is the name of Yahweh. From a fifteenth-century Florentine liturgical book.

John (a treasure trove of number mysticism), the victorious lamb has 7 horns; 7 seals are opened, and letters sent to the 7 churches; finally, 7 trumpets are blown to usher in the terrible Day of Judgment.

Medieval exegetes subsequently discovered many important features in the number 7. As the number of perfection, it is the day of God's rest but also points to the passing of time, since eternity begins only with Christ's resurrection on the eighth day. The 7 gifts of the Holy Spirit are counterbalanced by the 7 deadly sins. The 7 sacraments were divided according to the classical practice, into the higher, spiritual triad of baptism, confirmation, and eucharist, and the practical quartet of penance, holy orders, marriage, and extreme unction. They thus perpetuate the image of 4 cardinal virtues and 3 theological ones, related to body and soul respectively. The Lord's Prayer could be interpreted as a triad turning to God and a quartet related to humankind. It is interesting to note that the first sura of the Quran, the "Fatiha," which, in the Islamic context, is comparable to the Lord's Prayer, has the same structure: 3 of its 7 verses are addressed to God and 4 mention humanity's petitions and needs. The Islamic profession of faith, "There is no deity save God, Muhammad is the messenger of God," consists of 7 words: *la ilaha illa Allah Muhammad rasul Allah.*

Given the importance of the 7 in Christian tradition it is not surprising that the Catholic mass is arranged in its 7 parts according to old numerological principles, as Hopper has elaborated in detail. And in the fourteenth century the Egyptian historian Maqrizi reported that the Christians in Egypt (the Copts) celebrate 7 major and 7 minor feasts in their church. The 7 joys and 7 sorrows of Mary fit well into the heptadic rhythm. Thus one finds in Renaissance music a number of motets with 7 voices, which are usually devoted to the Virgin or else allude to the 7 gifts of the Holy Spirit.

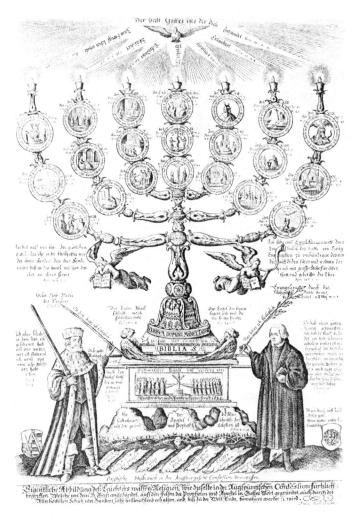

"True likeness of the candelabrum of the True Religion as it was recently understood in the Augsburg Confession." Copperplate engraving by Paul Fürst, *ca.* 1656.

Indeed, the number 7 was so central in medieval Christian thought that John of Salisbury in the twelfth century composed a book called *De septem septenis*, in which he discussed the 7 groups of things that manifest themselves in a heptad of diverse forms, beginning with the 7 kinds of erudition, then the 7 liberal arts, the 7 gifts of the Holy Spirit, the 7 degrees of contemplation, and so on to the 7 basic principles of philosophy. As 7 was considered the number of universality, medieval thinkers wondered whether the 7 churches mentioned in the Revelation were meant to point to the universality of the church or whether one should invert the logic and argue that, since there are 7 churches, 7 planets, and 7 spheres, 7 had to be a universal number.

Thanks to the transmission of Pythagorean doctrines and the mystical speculations attributed to Hermes Trismegistos, the medieval Christian tradition was based not only on biblical allusions but also on the classical heritage. In ancient Greece, 7 had an important place through its connection with Apollo and Athena. Singing swans circled around the island of

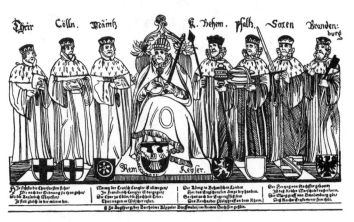

The German emperor with the 7 electors in order of their rank. Woodcut by Bartholomäus Käppeler, Augsburg, seventeenth century.

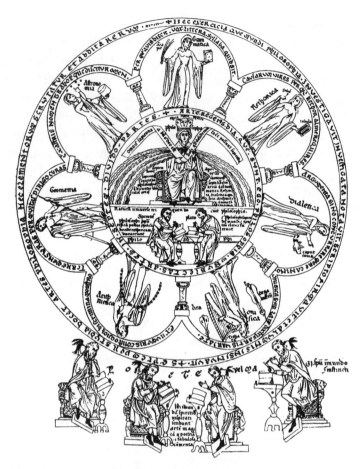

The 7 Liberal Arts as known from classical antiquity: Grammar (attribute: the rod), Rhetoric (the slate), Dialectics (the serpent), Music (instruments), Arithmetic (with counting beads), Geometry (compasses), and Astronomy (pointing to the stars). In the center, Queen Philosophy. From the *Hortus deliciarum* (1175–1185).

Delos 7 times before Leto gave birth to the radiant Apollo, and the birth itself took place on the seventh (or ninth) day, which is therefore dedicated to him. The lyre the god played had 7 (sometimes 9) strings. It has even been speculated that Apollo, who stayed with the Hyperboreans for 7 months, might have been connected with the 7 months of the cold season, but this is rather tenuous. The relationship to Athena was even stronger: since 7 is a prime number, neither producing nor produced (i.e., it is indivisible and has no product in the first decade), it seemed especially fitting for Athena, the virgin who sprang from the head of Zeus. This was expressed well by Philolaus, who wrote in the fifth century B.C.E. that 7 is "comparable to the goddess Athena, the leader and ruler of all things, eternal as a deity, steady, immobile, similar only to itself, different from all others." The fact that 7 is not capable of procreation was taken over in Jewish mysticism and applied to the Sabbath, the seventh day, on which humankind is supposed to rest and not to create anything. In ancient Greece Pythagorean ideas concerning the 7 were elaborated in particular by Nicomachus of Gerasa, who spoke of the relations between the 7 planets, the 7 tones in an octave, the seven musical keys, and the 7 Greek vowels.

Notwithstanding the association with the virgin goddess Athena, the 7 has also been connected with the patriarchate, a view supported by the use of 7 in ancient Rome, with its patriarchal social structure. Rome is built on 7 hills (which, incidentally, is claimed for many other places too). Seven was the auspicious number of the Roman villager and guaranteed the rule of the constantly expanding state. The expression *Roma septemgeminata* (sevenfold Rome) points to this belief. Even in the circensian games the 7 was used as a lucky number, for Rome was said to be founded in the first year of the seventh Olympic games. This may well be a legend, but there were dozens of institutions connected with the heptad,

The 7 steps leading to the Temple of Wisdom. The temple is hidden inside the "Mountain of the Initiates." This, as Jung writes in *Alchemy and Psychology*, points "to the fact that the philosophers' stone lies in the earth and has to be extracted and cleaned." In the corners, the 4 elements. The zodiac that rises above the mountain represents time. In front of the steps at the lower right, a deluded man (with his eyes bound); in the lower left, the scientist, keen on studying nature. Copperplate engraving from Michel Spacherus, *Cabbala speculum artis naturae in alchymia* (1654).

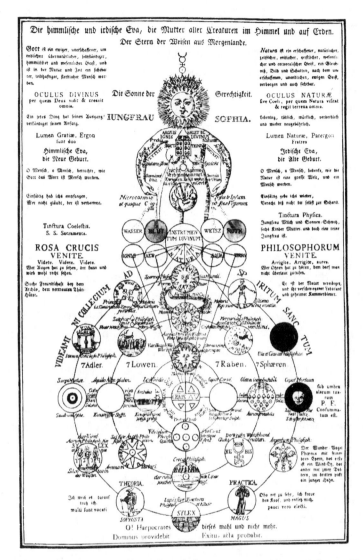

Seven as the fundamental number of the Rosicrucians. From *Geheime Figuren der Rosenkreuzer* (Altona 1785–1788). The main text of the Rosicrucians, *Chymische Hochzeit Christiani Rosencreutz* (1616), is completely organized according to the number 7.

such as the *septemviratus,* an assembly of 7 men who played a key role in the administration of the state (and one should remember that even in our time 7 persons are required to found a sodality according to German legal prescriptions.) The Roman heptads were more or less this-wordly, and there is no divine heptad found in the Roman pantheon. But the importance of the 7 was well expressed in early Christianity when the Roman church father Tertullian calls God the "septemplex spiritus qui in tenebris lucebat, sanctus semper" (the sevenfold, ever holy spirit who radiates in the darkness).

By early Christian times, however, a good number of mystery cults had also found their way into the Roman Empire, and many of them utilized the old heptadic division into divine and spiritual beings. In late antiquity in Egypt one finds the 7 carriers of the scepter who were connected with the 7 planets and days. Their names reached France via Rome and were then taken over into the Anglo-Saxon and Germanic traditions: thus, our days of the week are still connected with the names of astral deities, beginning with the sun (Sunday) and moon (Monday). Mars is reflected in the French *mardi,* the equivalent of Tuesday or *Dienstag* in English and German respectively, since Ziu = Thiu corresponded approximately to Mars. Then follows Mercury (still *mercredi* in French, the English Wednesday, that is, Wodan's day), Jupiter (in French, *jeudi;* in English and German Thursday, *Donnerstag,* from Tunar, Donar), Venus (French *vendredi,* English and German Friday, *Freitag,* after the goddess Freya), and finally Saturday, which bears the name of Saturn. In the Christian gnostic tradition, the ancient rule of deities over the planetary spheres was transformed into a relation between the 7 lower spheres and the demonic powers (which replaced the pre-Christian deities). Such ideas are found with the Ophites who, like other gnostics, regarded these 7 archons (i.e., the "planets") as leaders of the material world; they in turn are ruled by a higher spirit who sometimes appears as feminine, such as the Ruaha, an evil female demon. In the Mandean gnosis all these spirits are evil, hence, 7 is a dangerous number for the Mandeans: as a Mandean scripture claims, 7 seducers have seduced all the children of Adam.

Even more important for the history of religion in Western Europe was the role of 7 in the deeply influential Mithras cult. Mithras is originally an Iranian sun god, but around the beginning of the common era, the cult connected with him, like a number of others, developed into a mystery religion. In the mysteries of Mithras, the soul was thought to rise through the 7 planetary spheres into the divine presence. This ascension was symbolically represented by 7 gates through which the adept had to pass, leaving one piece of clothing at each gate to symbolize the shedding of one human quality after the other. This rite goes back to ancient Babylonia, where it was told that Ishtar, when traveling to the netherworld, had to leave one garment at each of the 7 gates. In the Mithraic mysteries, adepts finally reached an eighth portal, the Gate of Light, where they stood naked, divested of all material qualities and ready to be reborn in the spiritual world. The rites of expiation and purification connected with this cult took place, as might be expected, on the seventh, fourteenth, twenty-first, and twenty-eighth days of the month. It may be that a children's game something like hopscotch contains a dim memory of this mystery cult, which came to Germany and Britain along with the Roman soldiers: in this game (which is still played), one leaps on one leg through a ladder-like figure drawn on the ground, and the last station in the eighth square is variously called heaven or hell.

The 7 degrees of initiation into the Mithras cult as well as the ancient ideas about human ascension through the starry spheres form the basis of the Christian concept of the 7 layers of purgatory, but more important, they prefigure the gener-

The "son of the earth" who, as the first human being, undertakes a journey through the 7 heavens; here represented on the exterior of a shaman's drum from the Jenissei. The outer circle has 7 small bays; the inner circle shows 7 protruding points and an eighth sign with a bird, which alludes, according to Eliade, to the number of its heavenly planes and the 7 sons of the sky god, respectively.

Mythical 7-branched tree on a Chinese clay brick of the early Han period (206 B.C.E.–8 C.E.). According to Hans Findeisen, this may well be the first representation of the Cosmic Tree with its 7 levels of branches, on which the shaman climbs up to the 7 heavens.

ally accepted concept of the 7 steps on the mystical path. It should be remembered, however, that a similar concept can also be encountered in Siberian shamanic cults: the cosmic pole on which the shaman rises often has 7 incisions, and the Samoyed shaman may lie unconscious for 7 days and 7 nights before he undertakes his task; he may also eat mushrooms with 7 spots and perform rites containing the 7 (or the 9).

The sevenfold path seems to be one of the most universal ideas in the world of religions, whether it is expressed by the Flemish mystic Ruysbroek speaking of the 7 steps in mystical ascent or by the Persian poet 'Attar in the late twelfth century telling about the journey of the soul birds through 7 valleys, or by the Iraqi Sufi Nuri describing (*ca.* 900) the 7 walls of the spiritual castle, or by St. Teresa of Avila visualizing the 7 interior castles. Behind all these descriptions of the path is the old idea of the ascent through the 7 planetary spheres.

As can be understood from the names 'Attar and Nuri just mentioned, Islam too gives the 7 an important role, because of its Semitic roots on the one hand and because of ancient Persian traditions on the other. Indeed, the exchange between the two cultural areas was lively and helped form a whole Persian Islamic cultural tradition. The Mithras cult, for example, came from Iran, and Herodotus tells that there were 7 tribes in Iran, each of them with its own fire temple. The division of the world by the Persians into 7 climata was also known, and the Persian national epic, Firdawsi's *Shahnama* (Book of Kings, written shortly after 1000 C.E.) tells of the 7 heroic acts of Rustam and of another heptad of feats performed by Isfandiyar. King Gurshasp is blessed with 7 sons, and Giv has to wander around in his quest for 7 years.

This role of the 7 is already attested in ancient Zoroastrianism, where the 6 Amesha Spentas, or guiding spirits, were complemented by Ahura Mazda, the deity of justice and goodness, so as to form a heptad. In the *Arda Viraf Nama*, a

Persian book describing the soul's ascent to the next world, Viraf journeys for 7 days before reaching his goal. Numerous customs in Iran use the number 7, beginning with the description of the human being as *haft andam*, with 7 limbs (i.e., 2 legs, 2 arms, belly, breast, and head). Especially during wedding ceremonies 7 people have to carry out special duties. (Likewise, in Pakistan 7 happily married women should make the first 7 cuts in the material for the bridal dress.) If someone falls ill, one piece of food should be collected from each of 7 houses and given to the ailing person, who, it is supposed, will thus recover. In modern times, a special religious meal arranged by women is prepared with food begged from 7 women named Fatima. Many recipes are based on 7 ingredients; *haft maghz*, "7 kernels," is a kind of marzipan, and there are 7 kinds of *turshu*, or pickled vegetables. For Nawruz, the Persian New Year that is celebrated on the vernal equinox, 7 items whose names begin with the letter *s* are displayed in the home. There are also 7 styles of calligraphy.

Early Islam too was aware of the importance of 7. According to the Quran, God created heaven and earth in 7 layers. The *tawaf*, the circumambulation of the Kaaba in Mecca during the pilgrimage, has to be performed 7 times, and also the

Turkish amulet formed from the names of the 7 Sleepers, in the center that of their dog. Nineteenth century.

running between the stations of Safa and Marwa, and at the end of the *hajj,* the devil is stoned near Mina with 3 volleys of 7 stones each. Tradition has preserved 7 "suspended" odes, the *muʿallaqat,* as treasures of pre-Islamic poetry in Arabia (although that may well have been a round number to begin with). According to a *hadith* (tradition) transmitted by Bukhari, there are 7 major sins. The 7 Sleepers, known from early Christian history, appear also in the Quran (18:21), and the names of these pious young men, "whose eighth was their dog," are used both in the Islamic tradition and that of the Eastern Orthodox church for amulets, often written in beautiful calligraphy. Sura 15:87 speaks of 7 *mathani,* "doubled ones," possibly referring to the 7 verses of the "Fatiha," which was revealed twice. Among the mysterious, unconnected letters of the Arabic alphabet that appear at the beginning of a number of quranic suras, the combination *ḥm* appears 7 times; this was later interpreted to mean *ḥabibi Muhammad,* "my beloved Muhammad." The word *salam* (peace) also occurs 7 times. An early *hadith* speaks of the 7 internal aspects of the Quran, which have been emphasized in later mystical writings and Shiite esoteric hermeneutics. There are also 7 canonical forms of Quran recitation, and at a certain station along the pilgrimage to Mecca, the pilgrim exclaims 7 times "Allahu akbar!" (God is greater [than everything]). The 7 letters of the Arabic alphabet that do not occur in the "Fatiha" later came to be used in magic or for the construction of magic squares.

Particularly active in numerological speculation were the Hurufis, a Muslim sect that appeared in Iran in the late fourteenth century. Convinced that everything was contained in the letters and their numerical values, they easily drew correlations between the heptads found in the Quran and the 7 parts of the human face, the hair, and the rest of the body.

The 7 was especially dear to the Sufis. Later Sufism speaks

of 7 *lata'if,* subtle points in the body upon which the mystics concentrate their spiritual power; these have a parallel in the *chakras* known from Indian mystical systems. By wandering, as it were, through the *lata'if,* the Sufis can reach higher and loftier levels of consciousness during their constant meditative prayers. These *lata'if* are seen in relation to the 7 essential qualities of God which, in turn, are manifested in the 7 major prophets from Adam to Muhammad. For this reason, heptads play an important role in visions: the mystic sees 7 candles, is invested with 7 spiritual robes, and so on. In one of the Sufi fraternities, the Tijaniyya, it is believed that the sevenfold repetition of a certain litany results in the prophet Muhammad himself attending the meeting of his followers, who bless him with this formula.

The mystical hierarchy in Islam is built on 7 degrees, the highest one being the *qutb* (pole), or axis of creation, and the lowest one, the 4000 hidden saints. Groups of 7 saints appear frequently; to mention only one example, the city of Marrakesh in Morocco is under the special protection of the 7. Parallel to these groups are heptads of virtuous ladies who appear as martyrs in Sufi history and folklore. (The notion of 7 martyrs is also known in Christian tradition.)

Popular parlance is replete with sayings that contain the number 7. In Iran the cat has 7 lives and carries her kittens 7 times to different places. The heavenly spheres are known as "7 mills," and the constellation Ursa Major as "7 thrones." "To do the work of 7 mullahs" means "to accomplish nothing," and where the German says that he or she is related to someone "through 7 bushels of peas" (*durch sieben Scheffel Erbsen*) to denote a very great distance, the Persian will say, before mentioning something negative or dangerous: "May 7 Qurans (or 7 mountains) be between it (and the listeners)."

Books and tales are often put together in heptads; the prime example of the literary use of the 7 is Nizami's epic

Haft paykar (The 7 pictures). In this work the Persian poet tells how the hero visits a different princess each day of the week, and the day, the relevant star, the star's color, fragrance, and other influences are all elaborated according to the rules of astrological wisdom. One may also think of the stories of the *Seven Wise Masters*, which became popular in Europe in the Middle Ages, or of Sindbad's 7 voyages. There are books called *Seven Oceans* (one of them, an introduction to the Persian rhetorical system, was translated into German by F. Rückert in 1827 and is still the best introduction to Persian figures of rhetoric). There are also treatises called *Seven Spheres* or *Seven Climata*, and such Persian works are not, as one might expect, geographical studies but rather treatises on different kinds of literature, the 7 in their titles expressing completeness.

However, the role of the 7 becomes most important in the doctrine of the Ismaili sect, also called Sevener Shia because the last imam (leader of the community) was the seventh in the chain of imams from Muhammad through 'Ali and his martyred son, Husayn. In this current, which produced a remarkable philosophical literature from the tenth century on, great heptads were formed, apparently under strong Hellenistic-gnostic influences. Enoch, Noah, Abraham, Isaac, Jacob, Moses, and Jesus are the 7 pillars of the world, that is, the 7 pillars of the House of Wisdom or the 7 shepherds. Revelation comes in 7 cyclical periods, and the seventh imam in the succession of the seventh, last, prophet, will inaugurate the resurrection. Ismaili philosophers recognized in the 7 letters of the divine word of creation, *kun fayakun* (written *knfykun*), "Be!, and it becomes," the principles out of which flow 7 primordial fountains; from the primordial mist come 7 heavens and then 7 earths. The 7 great prophets correspond to the 7 spheres, the 7 imams who appear in each prophet's period correspond to the 7 earths; the 12 substitutes of each prophet correspond to the 12 signs of the zodiac, and the 12 *hujjas* (proofs) for each imam correspond to the 12 "islands." The heptagonal fountain in the new Ismaili Center in South Kensington, London, symbolizes these ideas in artistic form.

The Brethren of Purity in Basra, who were close to Ismaili ideas, attributed humans with 7 vegetative and 7 spiritual faculties and claimed that creation unfolds itself from God through the First Intellect to man in 7 steps. Nasir-i Khusraw, the eleventh-century Ismaili philosopher, divided the entire universe into heptads of spiritual and material things. The Druze community, which grew out of the Ismailis, has a sevenfold set of commandments. Yet another tradition, the Babi-Bahai faith that emerged out of Shiite Islam, also stresses the importance of the 7: the Bab ("gate") who founded this movement in Iran in the early nineteenth century is called "the man of 7 letters," since his name, ʿAli Muhammad, is written with 7 letters (ʿlymhmd).

The 7 also occurs frequently in India. In fact, along with the 3, 7 is the most important number in the Vedas. It is especially connected with Agni, the god of fire, who has 7 wives, mothers, or sisters as well as 7 flames, beams, or tongues, and songs devoted to him are sevenfold. The sun god owns 7 horses that draw his chariot through the skies. The *Rgveda* mentions 7 stars and 7 streams of heavenly soma, the drink of gods. Indra, the god of storm and rain, is the "7-slayer," and there are 7 parts of the world, 7 seasons, and 7 fortresses in heaven; the ocean has 7 depths and one speaks of 7 concentric continents. The predilection of the ancient Indians for 3 and 7, moreover, results in combinations of the 2 numbers in the *Rgveda*: 7 rivers grow into 21, and even the cow has 3 × 7 names. "Whatever exists in the realm of deities and in the realm of the self—all that is what he will obtain through the hymn with 7 verses" says the Satapata Brahmana (IX 5,2,8). It is natural that such an important number should

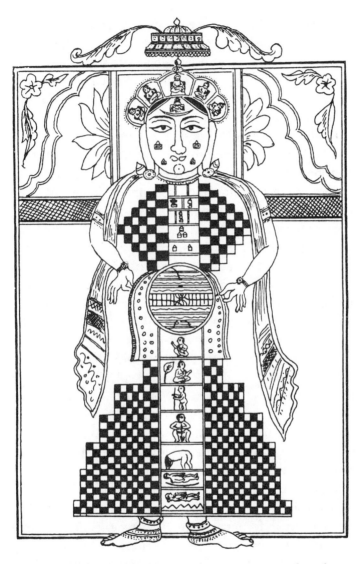

Cosmographic diagram of the Jains. The universe is imagined as a human-shaped being. The gods live in heavenly palaces in the head and crown while the lower "body" contains a 7-storey netherworld.

also be used in ritual—hence, for example, the 7 steps around the fire performed in the marriage rite.

It is clear that such ancient Indian concepts were taken over in Buddhism. The newborn Buddha takes 7 strides to express the fact that this will be his last birth. He seeks salvation for 7 years, and circumambulates the Bodhi tree 7 times before sitting down in meditation under it. The Buddhist paradise has 7 terraces, and 7 religious works will bear fruit for the believer in this life. In the scientific realm, 7 is at the basis of the incredible accumulation of numbers in later Buddhist literature, going from 7 atoms to 7 utmost minute particles to 7 minute particles to 7 particles, and so forth until one finally reaches the mile, which consists of $10^3 \times 4 \times 2 = 7^{10}$ units.

According to Japanese popular belief 7 deities bring good luck. Among the Germanic peoples 7 was not a sacred number, but it does occur as a round number in the *Older Edda* where it is written: "7 days we trotted through cold country, 7 more days over the sea, and in the third 7 we journeyed across dry steppes." It is likely that 7 reached the Germanic areas through the Mithraic cults on the one hand and Christian missionary activities on the other and then took the place of the indigenous 9 in various connections.

The longstanding role of 7 as a magic number is evident from its place in magical and especially alchemical processes. For this reason, the 7 Seals of Solomon were frequently used in Islamic magic, and the strange, mysterious name Abraxas, which consists of 7 letters, also plays an important role there. Just as effective formulas were often repeated 3 or 7 times, even more so were important procedures in the alchemist's laboratory repeated. Distillations, for instance, usually had to be performed 7 times.

The 7 is important in ancient medicine as well: in the tradition of Hippocrates it is said that the number 7 rules the ailments and everything that is destructible in the body. The physicians of antiquity knew that painful ailments lasted for 7 days or a multiple of 7. The Pythagoreans called 7 a "crisis," and considered all days that could be divided by 7 as critical, including the seventh and fourteenth days of the month. Such ideas were preserved in folk medicine: in Germany it used to be believed that pigs would not contract hog-cholera if they were treated for 7 days with drinking and bathing water containing asphodel. In German folklore, the 7 sisters appear as fever demons, and healing takes 7 or 9 days or weeks. To get rid of certain ailments such as gout, it was recommended to recite formulas with 7 and 3, as in this medieval German prayer that was to be repeated 3 times:

> Hast du eines von den 77 Gichtern, so gesegne ich es dir.
> Es gehen drei heilige Männer herfür:
> Der erst ist Gott der Vater, der zweite Gott der Sohn,
> der dritte ist das Marienkind, das dir deine 77 Gichter wegnimmt.
>
> (If you have got one of the 77 gouts I bless you.
> There come 3 holy men,
> the first is God the Father, the second God the Son,
> the third is the child of Mary who takes away your 77 gouts.)

Such sevenfold repetitions can be enlarged to absurd formulas, as in the example cited by Trachtenberg in *Jewish Magic and Superstition:* to cure a tertian fever, "Take 7 prickles from 7 palmtrees, 7 chips from 7 beams, 7 nails from 7 bridges, 7 ashes from 7 ovens, 7 scoops of earth from 7 door sockets, 7 pieces of pitch from 7 ships, 7 handfuls of cumin, and 7 hairs from the beard of an old dog, and tie them to the neck-hole of the shirt with a white twisted cord." Seven years are generally required to be rescued from ailments caused by witchcraft, and apparitions such as the White Lady, which are connected with old, spooky castles, tend to reappear every 7 years.

There are also more harmless uses of the magic 7. In Bavaria, the yarn spun by girls under the age of 7 was considered especially precious (*Siebenjahresgarn*), for a child at age 7 can see hidden treasures. It was even believed that a black rooster, after reaching the age of 7, would lay an egg containing a dragon.

Because of its wide use, 7 could become an indefinite round number meaning "many." The exegete Donatius Tyconius states in his *Seven Rules for the Interpretation of the Holy Writ* that the stereotype number should be regarded as symbolic: that is why one finds frequent use of the 7 winds, the 7 seas, the 7 climata, the 7 eras, or, in the Oriental tradition, the 7 deserts, and 7 wisest men are paralleled by the 7 wonders of the world. Poets can be credited with 7 birthplaces or 7 burial sites (Homer in ancient Greece, Yunus Emre in medieval Anatolia). The numerous local names formed with 7 also express a multitude: Siebenbürgen (Sevenoaks), Sacelele ("7 villages" in Romania), or Yedikule ("7 towers," the fortress around the ancient part of Istanbul). Similarly, the hydra has 7 heads and the shield of Ajax 7 layers, and one should not forget the 7 dwarfs behind the 7 mountains and other groups in fairy tales, be it 7 ravens, 7 little goats, or 7 riders on white horses. Poets have often used the span of 7 years to express a long time, as in Hermann Löns's song:

> Rosemarie, Rosemarie, sieben Jahre mein Herz nach dir schrie. . . .
>
> (Rosemary, my heart was crying for you for 7 years)

or Börries von Münchhausen's *Ballad of the Nettlebush*, where the nettle stood for 7 years along the roadside to remind the king of his broken promise to his beloved. Even children's rhymes like:

> Wer will guten Kuchen backen,
> der muß haben sieben Sachen. . . .
>
> (If you want to bake a good cake you need 7 things)

belong in this group. It suffices to cast just a cursory glance at the titles of German or English novels or collections of articles to recognize the importance of the heptad (besides the triad), whether one thinks of Gottfried Keller's *Fähnlein der 7 Aufrechten* or Agnes Miegel's *Fahrt der 7 Ordensbrüder*, both of which center around 7 heroic and faithful men. And while the dancer and choreographer Mary Wigman created the "Seven Dances of Life" in 1921, Nietzsche wrote of *Siebente Einsamkeit*, the seventh loneliness, which was quite in tune with the seventh grade of the mystical experience.

And finally there is the Arabic proverb that speaks of 7 things of which one never gets too much: "bread offered in kindness, meat of lamb, cool water, soft garments, beautiful fragrance, a comfortable bed, and the view of everything that is beautiful."

THE AUSPICIOUS
NUMBER

○　○　○　○　○　○　○　**8**

In antiquity, the number 8 was considered interesting for purely mathematical reasons: the mathematicians of ancient Greece discovered that every odd number above 1, when squared, results in a multiple of 8 plus 1, thus $5^2 = 25 = (3 \times 8) + 1$, expressed by the formula $u^2 = (n \times 8) + 1$. They also discovered that all squares of odd numbers above 1 differed from each other by a multiple of 8, thus $9^2 - 7^2 = 81 - 49 = 32$, or 4×8. In architectural design, the octagon is the first form to serve as the transition from the square to the circle, which is important in the construction of domes.

But 8 is more than an interesting mathematical item. Already in antiquity it was regarded as a remarkable, lucky number: it was thought that beyond the 7 spheres of the planets the eighth sphere, that of the fixed stars, was located. As a "number of gods" it is found as early as in ancient Babylon. In Babylonian temples, the deity resided in a dark room in the eighth storey, and it may be that the association of 8 with Paradise is based on this custom. In the Mithraic mysteries one also finds a mysterious eighth gate beyond the 7 main gates; by passing this "mountain of transsubstantiation," the adept would be able to return to the luminous spiritual homeland after death.

The connection of 8 with Paradise continues through the ages: the Muslims believe that there are 7 hells and 8 paradises, since God's mercy is greater than his wrath. This idea is suggested by the title *Hasht bihisht*, "8 Paradises," found several times in Persian literature, and may be reflected in the fourfold and eightfold division of gardens common in Iran and Muslim India. The link seems especially likely when such gardens surround a mausoleum, since the Quran promises a "garden under which rivers flow," and thus the garden with its 8 parts may prefigure heavenly beatitude. The custom of dividing gardens into 8 parts carries over to one of the most famous works of medieval Persian literature, Sa'di's *Gulistan* (Rose garden), which is divided into 8 chapters. Later imitations of this book follow the same arrangement. And along with the 8 paradises, Islamic mythology posits 8 angels engaged in carrying the divine Throne.

Another root of the important, and lucky, aspect of the number 8 may be that in the ancient Near East, among the Elamites, there was a Venus year of 8 months, which was expressed symbolically by an 8-pointed star. Like the pentagram, this star was a symbol of Ishtar, goddess of love and

Calligraphic design made from the names of the 10 companions of the prophet Muhammad who were promised paradise. The octagonal design alludes to paradise. Egypt, eighteenth century.

fertility. Taken over into early Christianity, it is found in the catacombs of St. Priscilla in Rome, accompanying a representation of the Virgin Mary. More generally, both the 8-pointed star and the octagon were taken over by the Jews and then the Christians as signs of good luck.

Eight appears as a second beginning, on a higher level, the fulfillment of what the heptad had prepared and completed. Therefore, it is the day of purification in the Jewish tradition. More important, it is the eighth day on which circumcision takes place: "And on the eighth day the flesh of his foreskin shall be circumcised" (Lev. 12:3).

It was convenient for Christian exegetes to take over this idea of regeneration: Christ's resurrection took place on the eighth day of the Passion, thus promising future glory and eternal life to the believer. In addition, the name of Jesus in Greek letters, *IHESOYS*, yields the numerical value of 888, a multiplication and hence strengthening of the sacred 8. Gernot of Rechersberg, a medieval German theologian, therefore says: "Octonarius primus in numeris cubicis est, aeternae beatitudinis nobis in anima et corpore stabilitatem simul at soliditatem designans" ("Eight, as the first perfect cube [2^3], imprints us in body and soul with the security of eternal beatitude").

For the Church fathers, circumcision, baptism, and resurrection were mysteriously connected, with all of them expressing entrance into the life of salvation. Baptism is, as Augustine says, the circumcision of the heart, and on the day of baptism, according to Cyril of Alexandria, "we are made similar to the resurrected Christ as we have died in spirit by means of baptism and have become companions in the resurrection." This connection with the eighth day, the day of circumcision and resurrection, is one reason for the octagonal shape of many medieval baptisteries: baptism promised the Christian divine grace and the beatitude of eternal life. In-

The octagonal shape of the baptistry (here an example from the fourteenth-century church of Santa Maria in Visso, Umbria) is a symbol of the new life after baptism. The importance of the paradisical number 8 is emphasized here by the 8 small columns at the corners. According to early Christian thought, Christ's resurrection constituted the eighth day of Creation, and baptism, prefigured in the Jewish circumcision on the eighth day, thus marks entry into eternal life.

deed, Dante rightly places the triumphant church in the eighth heaven in his *Divine Comedy.*

In the Sermon on the Mount, Christ mentions 8 beatitudes, but in fact, it seems that the eightfold division of the path that leads to eternal bliss is a rather universal concept. Just as the Buddha teaches the noble eightfold path leading to cosmic equilibrium, the basic rules for the aspiring Sufi in Islam are also expressed in the 8 sentences of the so-called Path of Junayd.

The idea of 8 as auspicious, which is reflected in the Buddha's teaching and the numerous uses of 8 and its multiples in Buddhism, are rooted in the Indian tradition, where the 8-

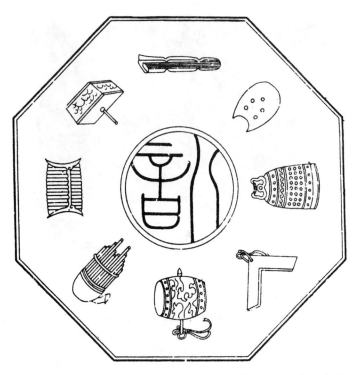

Eight Chinese musical instruments. In ancient China one often finds 8 attributes of an art, science or religion: 8 symbols of Buddhism, 8 emblems of a scholar, or of Confucianism, 8 symbols of the "immortals" in Taoism.

petaled lotus represents such luck and beatitude. In China too the 8 is highly esteemed, not only through the 8 symbols of Buddhism but also through the 8 precious items of Confucianism. By dividing the 4, the number of the well-organized created world—the doubling of even numbers is a frequently used device to guarantee an even greater power—one finds 8 winds, 8 pillars of heaven, and 8 gates for the rain clouds. As an even number, 8 is in most traditions connected with the feminine gender; yet, in China it determines the life of man: he has his milk teeth at the age of 8 months, loses them at 8 years, reaches puberty in 2 × 8 years and loses sexual

The 8 Immortals in China: Chang Ko-lao on the donkey, Chung-li Ch'üan with the fan, Han Hsiang with the flute, Ho Hsien-ku with the magic lotus flower, Lan Ts'ai ho with the basket of flowers, Li T'ieh-kuai with gourd and bat, Lü Tung pin with the sword, and Ts'ao Kuo-chiu with castanets. Ancient woodcuts.

strength at the age of 64, that is, 8 × 8. Furthermore, a scholar is distinguished by 8 special symbols. It is likely that the auspicious character of 8 as number of good fortune and perfection may underlie the 8 × 8 = 64 configurations of the *I Ching*.

In many cases, as just mentioned, 8 seems to express an empowered 4, and the 8 legs of the German god Odin's horse may express his swiftness. But it is also possible that these 8 legs reflect ancient nordic concepts of the division of the horizon (doubling of the cardinal points) and the year. In the ancient Germanic tradition, the 8-spoked wheel was often used to symbolize the year.

Eight can be used as a round number, as in the German expression *acht Tage* (8 days), which really means a week with

Left: Odin on his 8-legged horse, Sleipnir. From one of the famous pictorial stones in Gotland (Stone of Ardre VIII, *ca.* 1000 c.e.). As Eliade states, the 8-legged horse is the quintessential shamanic horse; it is found in Siberia and elsewhere, always in connection with ecstatic experiences. In Nordic belief, the 8 is usually connected with witchcraft and magic. Thus, from Odin's ring, Draupnir, 8 rings trickle down, and Geirröd kept the disguised Odin "between two fires" for eight days. *Right.* Tibetan magic diagram. The 8 heads are built on square and circle. They may be symbols of the 8 winds, or the 8 pillars of the sky, or 8 mountains, 8 gates, etc., of Sino-Tibetan tradition. Old woodcut.

its 7 days. The same is true of "huitjours" (8 days) in French. Similarly in the Japanese tradition: 8 comprises "infinite numbers," and in antiquity, Japan itself, consisting of innumerable islands, was simply called "great 8 islands." But strangely enough, despite its lucky qualities, 8 has barely played a role in popular religion, and magic. Even so, there was a remarkable number of couples in Germany who rushed to the town halls on 8.8.88 to get married on this auspicious day!

THE MAGNIFIED
SACRED 3

○ ○ ○ ○ ○ ○ ○ 9

The number 9 can be variously interpreted. At times, the purely negative aspect is stressed, thus by Petrus Bungus, who equates 9 with pain and sadness and points out that the ninth psalm contains a prediction of the Antichrist. The 9 is again connected with suffering because of the fact that Christ died at the ninth hour of the day (from sunrise, i.e., 3 P.M.). This hour was the *none,* subsequently marked by a special monastic devotion, while the word itself has become our *noon.*

Another type of exegesis, known from classical antiquity, emphasizes the near-perfection of the 9: Troy was besieged for 9 years, and Odysseus traveled for the same amount of time. This type of interpretation is typical of much of medieval Christian exegesis, based on the parable of the lost piece of silver (Luke 15:8–10). The 9 orders of angels found in such exegesis (and in Dante) are interpreted as reflections of the perfect 3, which can be completed by the all-embracing divine Unity to form the perfect 10. Another aspect of this "heavenly" interpretation of 9 can be derived from its role as 8 + 1, beatitude augmented and enhanced.

In fact, the number 9 does not hold a very prominent place in the Semitic, and following it the Judeo-Christian, world. Rather, its importance is connected with the Indo-Germanic

The 9 steps leading to the Heavenly City, as seen in the *Liber de ascensu* by the Catalan mystic Ramon Llull (1235–1316), printed in Valencia in 1512. In his early writings, Llull had taken over much material from Islamic mystical texts, but he was stoned by Muslims in Tunis while attempting to convert them.

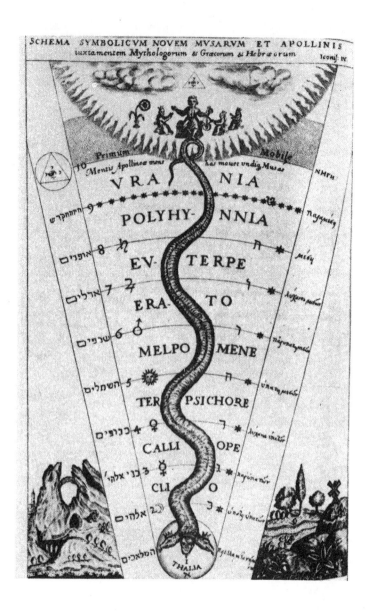

166

and Central Asian (Turco-Mongolian) traditions, and it seems to be typical of civilizations in the northern parts of the world. It has even been speculated that this is a result of the experience of 3 very cold winter months in contrast to the positive summer months, but this seems difficult to maintain. In many of these traditions, 9 is connected with the spheres, and the highest, ninth heaven is located beyond the 7 planetary spheres and the upper vault of the sky that contains the fixed stars. Persian and Turkish traditions, and book titles, often speak of *nuh sipihr* (nine skies). Mircea Eliade has pointed out that this may be a trebling of the ancient cosmic regions, to which also belong the 9 branches of the cosmic tree. That might explain why 9-storeyed pagodas were popular in China.

A similar role for the 9 can be observed in the philosophy of the Muslim Brethren of Purity with its 9 states of existence: one Creator, 2 kinds of intellects, 3 souls, 4 kinds of matter, 5 kinds of nature, the corporeal world determined by the 6 directions, the 7 planetary spheres, the 2×4 elements, and finally, the 3×3 states of the animal, vegetable, and mineral kingdoms.

In some civilizations, 9 was seen in connection with the

◀

The Harmony of the Spheres (*facing page*). The 9 spheres descend from heaven under the image of a tricephalic serpent and terminate on the earth. Each sphere has a corresponding muse: Thalia (comedy) belongs to the earth, Clio (history) to the Moon, Calliope (epic poetry) to Mercury, Terpsichore (choral song, dance) to Venus, Melpomene (tragedy) to the Sun, Erato (love lyrics) to Mars, Euterpe (music) to Jupiter, Polyhymnia (sacred song) to Saturn, and Urania (astronomy) to the heaven of the fixed stars. Above the heaven of the fixed stars, Apollo, the "leader of the Muses" (*musagetes*), produces the harmony of the spheres. In his hand, instead of the classical lyre, he holds a viola da gamba, a baroque string instrument. Above Apollo is the triangular symbol of the Trinity. Copperplate engraving from Athanasius Kircher, *Obeliscus Pamphilius* (Rome, 1650).

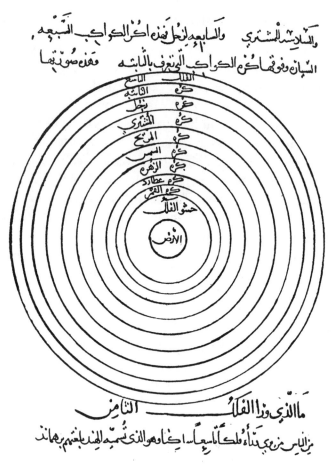

According to Islamic cosmology, the universe is built of 9 spheres. Next to Earth is the lunar sphere, above which are located the spheres of Mercury and Venus. The sphere of the Sun occupies the central place among the 7 spheres of the planets and is therefore often called "Center of the Universe." The last 3 planetary spheres are those of Mars, Jupiter, and Saturn. Then follows the eighth sphere, that of the fixed stars. Up to this point, the structure is the same as that of Ptolemy, but the atronomer Thabit ibn Qurra from Harran (d. 901) added a ninth sphere in order to be able to explain what he considered to be the "perturbation of the equinoxes," and most Muslim astronomers followed him. This ninth sphere is usually called *falak al-aflak*, the "sphere of spheres," and it is believed that it does not contain any stars.

lunar phases. Thus in ancient Mexico, the cycle went from the ascending ash-moon, the crescent, the first quarter, and the third octant to the full moon and then back to the descending ash-moon. On the basis of these periodic movements, the whole cosmos was then divided into various layers: 9 layers of a netherworld and 13 heavenly layers, with the earth located as a kind of intermediate zone. These 23 layers, plus the sacred number 5, connected with the ages of the world, yielded 28, which is the number of the moon. Thus, in Mexico 9 became a symbol of the netherworld, the earth, and—contrary to the traditions in Europe and Asia—the female.

The 9 rivers that appear in the lowest layer of the netherworld according to ancient Mexican mythology are found as well in ancient China, but here the 9 is the typical male number, an empowered 3. The 9 rivers of the Chinese netherworld manifested themselves as a 9-headed dragon; this was slain by the mythical hero Yu, to whom the turtle then brought the first magic square, engraved on its back. This square, as we have seen (pp. 29–35) is divided in 9 parts surrounding the central 5, and thus the whole world was considered to consist of groups of 9: the sky has 9 areas; the earth, 9 countries; each country, 9 mountain ranges; each mountain range, 9 passes, and the ocean has 9 islands. The city of Beijing, built millennia ago with the help of astrologers, consists of a center and 8 streets that lead there, so that it too repeats the ninefold structure.

The division of areas into 8 parts around a central ninth point is known from other civilizations as well. Another reflection of this ideal form is the position of the pins in certain bowling games, and when the skittle ball hits all 9 simultaneously, this is the supreme achievement, as indicated by the German expression *alle Neune*, "all 9 [pins]," meaning "super!"

Although in the Chinese tradition 9 is mainly connected

Phoenix with 9 chicks, a symbol of good fortune in China. As the first power of 3, 9 is a very strong *yang* number. The phoenix (*feng huang*) is considered by some as symbolizing the union of the male, *feng*, and the female, *huang*, elements; others regard him as the quintessential mythological representative of masculine power.

with heaven and cosmological problems, it is also identified with the 9 openings of the human body and 9 kinds of harmony; even the mythical moon-fox is supposed to have 9 tails. And in order to emphasize the importance of the wise master Lao-tzu, he was said to have been born 9 × 9 years after his mother conceived him.

Like the Chinese, the Mongols and Turkic peoples have been very fond of the number 9. One of the leading tribes of the Turks were known as the *Tokuz Oguz*, the 9 Oghuz. The Great Khan of the Mongols, before whom people had to prostrate themselves 9 times, was attended by 9 standards or, in the case of Genghis Khan, 9 yak tails. Therefore an eleventh-century Turkish poet from Central Asia, Yusuf Khass Hajib, who wrote the epic *Kutadgu bilig* ("The Knowledge of How to

The millennial 9-tailed fox of East Asia. After reaching that age he can transform himself into a young girl to seduce men. The monkey Songaku at the right side of this woodcut is only a hundred years old but is able to produce hundreds of monkeys from his hairs, which he blows away. Woodcut by the Japanese artist Hokusai (1760–1849).

Become Happy"), compared the sunrise to the appearance of the ruler before whom 9 gold-colored banners were carried (yellow was the favorite royal color among Turkic peoples). The Turks speak of 9 spheres "for there is nothing beyond 9." However, this number also played an even more important role: it was customary to give presents in groups of 9, as medieval Arabic sources tell, and *tokuz* (9) could simply mean "a present," since presents—even in Moghul India—were supposed to consist of 9 parts. The rulers in the Tatar khanate of the Crimea even received presents of $9 \times 9 = 81$ pieces! In Moghul India, where many Turkic customs continued, the chronicler Abu'l Fazl records that an officer of the rank of *tarkhan* was not liable to be punished provided the number of his crimes did not exceed 9. Among the Siberian Yakut, meanwhile, shamans used to place 9 innocent boys and 9 innocent girls beside them before beginning their magical rites.

Given these varied associations, it is understandable that 9 could develop into a round number in Turkish. "When the Wednesdays of 9 months come together," for example, means that one's work has increased so much that it can barely be completed. Something that lies "under 9 knots" is well kept. Someone chased out of 9 villages is a poor wretch who is thrown out everywhere, and a person, or animal, called *dokuz babali*, "with 9 fathers" is simply a bastard. But if someone is called "the 1 rod for 9 blind men," he is certainly the only true helper in times of distress and anguish.

The 9 is similarly prominent in the Anglo-Saxon-Germanic world. The Celtic Cymrians, the first inhabitants of Wales, used it in practical life as well as in legal affairs. Nine steps are used to measure distance: thus, a fire may be lit 9 steps away from a house, and a dog that has bitten someone may be killed 9 steps away from the house of his owner. Only when 9 persons get together to attack someone is it considered a real attack. The same is true in Germanic law, where 9 steps again appear as a legal measure, and tributes consist of 9 items (such as eggs). Nine days is often a legal period: in case of a divorce the wife had to leave her former husband's house after 9 days; the same was true after the husband's death. The ownership of real estate also ended in the ninth generation; however, it is not clear whether the custom to lease real estate or graves for 99 or 999 years is part of this old legal aspect of the 9 or is done to avoid the perfect round numbers 100 and 1000.

The importance of 9 is preserved in some English, and to a lesser extent, German sayings. Thus it takes 9 tailors to make one man; 99 tiny tailors occur in German fairy tales, and when someone screams the threat "Neun wie dich fress' ich zum Frühstück!" (I can easily devour 9 like you for breakfast), the miserable state of the poor addressee can be easily imagined. A stitch in time saves 9, and when someone appears most decoratively outfitted, he or she is "rigged to the nines." To be perfectly happy is to be on Cloud 9 and not, as in modern German, in Seventh Heaven—a typical case of the replacement of an old Germanic 9 by the 7.

The number 9 appears frequently in Celtic and Germanic tales and myths. King Arthur had the ninth part of his father's strength; 9 kings paid homage to him, and he had 9 butlers; he was also imprisoned 3 × 3 days. There is a story of one Germanic hero who could hold his breath under water for 9 days and nights and could do without sleep for the same span of time; his spear was as strong as 9 spears of others. Folktales recount the birth of ninelings, and when a man could boast of 9 children, it was proof of his potency. The Pied Piper of Hamelin charmed the children, it is said, with the ninth tune of his flute. Everyone knows that a cat has 9 lives, but the expression "cat o' 9 tails" is used in a negative sense for the scourge. It is also claimed that cats can transform themselves into witches once they reach the age of 9 years, and during the *Walpurgisnacht*, the night of 1 May, the witches are supposed to travel to their meeting place on the Blocksberg (in the Harz mountains of Germany) in chariots drawn by 99 (mainly black) cats.

The woman seer in the Norse epic, the *Völuspa*, sings:

> Nine worlds I know,
> Nine woods I know, of the mighty central tree
> in the dust of the earth.

And Shakespeare's "weird sisters" repeat:

> Thrice to thine, and thrice to mine,
> And thrice again, to make up nine.
> Peace! the charm's wound up.

In German mythology, Odin hangs on the tree for 9 days and nights and during this time learns 9 songs. The dangerous Fenriswolf is fettered by 3 chains and 6 other items, and

Heimdallr, the wisest of the gods, has 9 mothers. Divine beings of a lower order often appear in nonads or multiples of nine, thus the German Valkyries, the 9 kinds of elves, and, in ancient Greece, the 9 muses.

Nine was also important in sacrificial rites among the Germanic peoples: either 9 things were sacrificed or, as in ancient Denmark, a great sacrificial feast was held during which 9 heads of each male species were offered. The ancient role of 9 as an important religious number has led to its appearance in all kinds of magic acts. In Thuringia girls would collect leftovers of 9 kinds of food and then would sit at midnight around the table, hoping that the spirit of their beloved ones would appear to them. Throughout Germany the belief that a newborn baby prepares itself during 9 days either for life or for death was rather widespread. If humans should transform themselves into animals, or vice versa, this usually happened on the ninth day; swan maidens and Valkyries were thought to change their identity after 9 years. This connection of the number 9 with ghosts and spirits made it necessary not to tell anyone of a vision for 9 days, and likewise, if one chanced to meet a ghost, this encounter should not be revealed for 9 days.

The magic number 9 also plays a role in the healing of certain ailments. In healing, a ritual act was often repeated 9 times; thus, a bewitched person had to count backward from 9 to 1, and 9 knots in a ribbon were thought to help in cases of a sprained foot or hand (this custom was known both in Scotland and in Germany). 9 kinds of ailments can be cured by a mixture of 9 special herbs, and in the area of Göttingen (Germany) one used to prepare, on Maundy Thursday, a soup called *Negenstärke*, "ninefold strength," which consisted of 9 different green vegetables. A *Neunkräutersegen*, a blessing uttered over 9 different herbs, is supposed to help against 9 demons and 9 kinds of poison; its power is enhanced when these herbs are gathered and bound into a bouquet on St. John's Day, 24 June. In some regions of Germany it was believed that thieves could be found out by placing a stool made of 9 different kinds of wood in a church. In Germany one used to make a so-called *Notfeuer* in times of great distress. For such a fire, 99 men had to bring kindling, and all endangered creatures of the area, humans and animals alike, had to run 3×3 times, through the flames. More enjoyable is the superstition that one's wishes will certainly be fulfilled provided one can count the same 9 stars on 9 successive clear nights.

In many old folk tales the hero has ninefold strength or has to perform 9 major tasks. Amusing in this connection is a belief reported from the canton of Uri in central Switzerland: one should feed a young steer for 9 years, in the first year with the milk of one cow, in the second year from 2 cows, and so on, until the animal finally drinks the milk of 9 cows. Then it should be led by a virgin across the mountain range and let loose, whereupon it will be able to overcome all the demons and goblins in the mountains.

Nine can also be used as a number of perfection, or final limit. This is evident in German words like *Neunmänner-werk*, "the work of 9 men," used for something unusually great and impressive, or *neunhändig*, "9-handed," for describing someone very skilled. Similarly, *neunäugig*, "9-eyed," means very shrewd and cunning, and a super-intelligent person is known as *neunmalklug*, "ninefold clever" (although this last is slightly deprecatory).

Giants and heroes in the Indo-Germanic tradition of Greece, Iran, and India appear as 9 cubits long; this is *nau-gaza* in Persian, but the same is also said of the biblical figure King Og of Basan. Nine also represents a number of greatness and perfection in the Indian term *naulakha*, something worth 900,000, a word used in fairy tales and folktales to designate

The god of love, Kandarpa (or Kama), holding a bow and arrow, rides on an elephant composed of 9 girls. Painting on paper from Bhubaneshvar, nineteenth century. As Jutta Jain-Neubauer explains, the story usually told to accompany such pictures is as follows: Krishna, longing for Radha, was wandering through the forests of Vrindavana. The *gopis* (cowgirls)—perhaps because of their infatuation with Krishna or else with the intention to wean him away from Radha and attract him to themselves—decided to produce an elephant with their bodies. Krishna, lost in thought, took the composite beast for a real elephant and mounted it. While he was calling Radha full of love and longing, the *gopis*, delighted that their trick had worked, suddenly decomposed the "elephant."

the most precious necklace or the most coveted piece of jewelry.

The dark side of the 9 appears in several Nordic, and especially Finnish myths: ailments, for example, can appear as 9 evil brothers or sisters. As this kind of idea is found predominantly in the northern areas, one suspects that these evil spirits personify the nine long dark months of the polar winter. There is a Finnish tale according to which Mount Kippumaki has 9 caves, each of them 9 cubits deep, in which magicians store humanity's pain and suffering, while on top of this mountain the Fury Hilta and her demon associates cook the plagues that are destined to overcome mankind. In Finnish lore, ailments, pain, and harmful animals are born to an old woman after a pregnancy of 30 summers and 30 winters; she then produces 9 sons, giving birth to them on a stone in the water. The 9 bad sons of a queen of the North in Estonian tales is probably a parallel idea. But as like can be cured by like, one can also conjure up the Finnish house god Tontu by circumambulating the kitchen 9 times, and he will put in order whatever the 3 darkest months of winter have done to the family.

The Indo-Germanic heritage with the special place of the 9 can be observed in ancient Greece. The river Styx in the netherworld has 9 twists, and feasts are prepared to honor Apollo in Delphi every ninth year. The period was later abbreviated to a 5-year term. In later antiquity a feast in honor of Zeus was celebrated on the Lycean mountain every ninth year. There were reportedly even human sacrifices, but the person who performed the ritual slaying was, as it is told, banished from the country for 9 years. During the feast of Dionysus at Patra, 9 men and 9 women celebrated the ritual. Apollo, accompanied by the 9 Muses, carried the lyre with 9—or as we saw earlier, 7—strings, and according to one version of the myth, the labor of his mother, Leto, lasted for 9 days and 9 nights, and Eileithya, who assisted the poor mother, received from her a necklace of 9 parts. One wonders if this "necklace" has any connection with the umbilical cord, which brings blood and nourishment to the embryo during the 9 months of its gestation.

The 9 Muses have inspired many thinkers and writers; that is why Herodotus divided his work in 9 parts to honor them, and in this connection, it is not too farfetched to think of the *Enneads*, the "9 books" of Plotinus, whose Neoplatonic philosophy became an important ingredient in the development of Jewish, Christian, and Islamic mystical thought. One is therefore also not surprised to learn that Plato, according to legend, died at the age of $9 \times 9 = 81$ years.

In Christian circles the 9 was generally connected with the concept of the Trinity. Dante's *Divine Comedy* is the best example for the use of this trinitarian symbolism. Beginning with the poetical form of the *terzine*, the 3-line stanza in which his work is composed, everything points to the Trinity which is, in turn, fiendishly distorted in the *Inferno*. Just as 3 is more comprehensive than 1, $3 \times 3 = 9$ can better express and realize humanity's relationship with God. The orders of angels are 9, and for Dante, this 9, again, is revealed in his beloved Beatrice, about whom he says in the *Vita nuova* (30: 26–27): "This number was her true self," that is, the reflection of the angelic world.

According to Roger Bacon, the ninth house of the horoscope refers to peregrination and travel, to religion, faith, and divinity. It is the house of worship of God, of wisdom, of books and scriptures and comes under the rule of Jupiter, generally known as "major fortune." Thus 9 can be considered, under certain circumstances, a lucky number. In the Swiss province of Aargau, for example, it was believed that the last 9 ears of corn that are gathered from a field at the end of harvesting will bring good luck to the one who finds them; they are called *Glückskorn*, "grain of good luck."

The old role of 9 as a measure of time is alluded to in a gypsy song that Lüttich has quoted in his comprehensive study on meaningful numbers:

> Hier im Wald am grünen Hage
> steh ich Armer schon neun Tage,
> will mein Liebchen einmal sehen:

Hier muß es vorübergehen.
Hätt' es Küsse mir versprochen,
stände gern ich hier neun Wochen,
würden jemals wir ein Paar,
stände ich hier auch neun Jahr!

(In the forest green I stay,
poor me, nine days, day by day,
just to see my true love dear—
I think she'll be passing here!
Had she promised me a kiss,
Nine weeks would I stay like this;
Would to marriage she agree,
Nine years seemed not long to me!)

And in a similar vein e. e. cummings sings:

for every mile the feet go
the heart goes nine. . . .

COMPLETENESS
AND PERFECTION

> This number was of old held high in honor for such is
> the number of the fingers by which we count.

Thus says the Roman poet Ovid, and it seems that in our tradition at least, the 10 fingers have served as the basis of the familiar decimal system. Indeed, as W. Hartner has shown, most of the counting systems in antiquity are based on it. For instance, in ancient Egypt, the sign of the lotus flower meant 1000 and that of the boat, 10,000. For the Pythagoreans, the importance of 10 was beyond doubt, for it was regarded as the all-embracing, all-limiting "mother." As the sum of the first four natural numbers $(1 + 2 + 3 + 4 = 10)$ it was associated with the primordial one existence, the polarity of manifestation, the threefold activity of the spirit, and the fourfold existence of matter as seen in the 4 elements. Thus the 10 contained everything, and in geometry it could be represented as an equilateral triangle:

The gnostic *anthropos* (man) as divine being carrying in himself the 4 elements and connected with the 10, which means "perfection" because it is the sum of the first four integers 1 + 2 + 3 + 4. Woodcut from Albertus Magnus, *Philosophia Naturalis* (1560).

In 10, multiplicity turns again, on a higher level, into unity, for 10 is the first step toward a new multiplicity that leads to yet another step that begins with 100, and so on. Mystically speaking, 1 and 10 are the same, as are 100 and 1000.

Since 10 represents unity emerging from multiplicity, Aristotle acknowledged 10 categories, and the same feeling is reflected in many traditions by grouping books or words of wisdom in tens: the early *Rgveda* in India with its 10 books, for example, or the 10 Commandments given to Moses. Buddhism, too, has 10 commandments, 5 for the monk and 5 for the layperson, and the oldest rule for Sufis in the history of Islamic mysticism, developed in the early eleventh century by Abu Saʿid-i Abuʾl-Khayr, consisted of 10 parts.

Jewish tradition has always been aware of the central role of 10. Not only were 10 Commandments given to Israel but, as the *Zohar* claims, the world was created in 10 words, for it is said in Genesis 1 not less than 10 times: "And God spoke." There are 10 generations between Adam and Noah. Only rarely does a negative aspect of 10 occur, as in the 10 plagues in Egypt. On 10 Tishri, the Jewish Day of Atonement, the confession of sins is repeated 10 times, and on Rosh Hashanah, the Jewish New Year, 10 biblical verses are read in groups of 10. The importance of 10 becomes even more obvious in the speculations of the Cabalists, who developed the concepts of the 10 *sefirot*, those 10 archetypal manifestations that are the basis of all existence, a world of essential divine entities that flow without a break and without a new beginning into the invisible and visible world of creation, as Gershom Scholem describes them. The form of the 10 *sefirot* was sometimes explained as *macroposopia*, the cosmic form of the primordial Adam, *adam qadmon*. There is no doubt that Pythagorean trends influenced the system which, however, was elaborated in a most fascinating way. The idea underlying the cabalistic system is that the unity beyond description un-

The 10 stages of human life, accompanied by a German doggerel that claims: 10 years, a child; 20 years, youth; 30 years, a man; 40 years, well done; 50 years, standstill; 60 years, old age begins; 70 years, an old man; 80 years, does not know anything; 90 years, children's laughing stock; 100 years, may God have mercy on you.

Printed as one-leaf publication by Abraham Bach of Augsburg. The stages are not so much determined by life expectancy as by the "perfect" character of the numbers 10 and 100.

folds into trinity. These are the 3 higher *sefirot: keter* (color white), *hokhmah* (color grey), and *binah* (color black). The fourth *sefirah, hesed* or *gedullah,* is connected with blue, the fifth one, *gevurah,* with red, and their combination appears in the ninth *sefirah, yesod,* in dark purple. The sixth *sefirah, tiferet* (color yellow) is related to the seventh, *netsah* (green) and the eighth, *hod* (color orange). These, in turn are connected with the fourth and fifth *sefirot,* and the final point is the tenth *sefirah, malkhut,* or the *Shekhinah.*

One could develop the entire cosmic structure and inner divine activity from the connections between the 10 *sefirot*, and the medieval Cabalists have done that by using gematric manipulations with letters and the numerical values of letters in an ingenious way. However, it would be unfair to dwell only on the high mystical aspects of the 10 and forget that this number has a very practical aspect to it, namely the tithe.

It is natural that Christian exegetes should have used the 10. They would ask: is not the Roman number X = 10 an allusion to the cross of Christ, to the first letter of his name, written *Xristos* in Greek letters, and also to the 10 Commandments? In addition, the letter *iota*, with which the name of Jesus begins, had the numerical value of 10. In allegorical exegesis of the Bible, 10 could be understood as pointing to the 3 persons of the Trinity and the 7 elements of life (heart, soul, and mind plus the 4 elements). One could also explain it as the sum of 9, the orders of angels, and 1, the human being, or as an allusion to Job's 10 children in whom the 7 gifts of the Holy Spirit (the sons) and the triad of faith, hope, and charity (the daughters) are personified. Other exegetes discovered in the 10 Commandments 3 commandments of piety toward God and 7 commandments concerning relations among human beings.

Ten plays a role in the Islamic tradition as well. Five inner senses correspond to the 5 outer ones, and Muhammad mentioned "the 10 who were promised paradise" among the earliest Muslims. Their names were often written on amulets in the form of an octagon (typical of the paradisiacal connotations of the shape). Following the Prophet's example, major mystical leaders surrounded themselves with 10 particularly faithful disciples, or so legend tells. Turkish scholars were delighted to find that Sultan Süleyman the Magnificent, the tenth Ottoman emperor, was born at the turn of the tenth century of the hegira and had 10 children. It was easy to

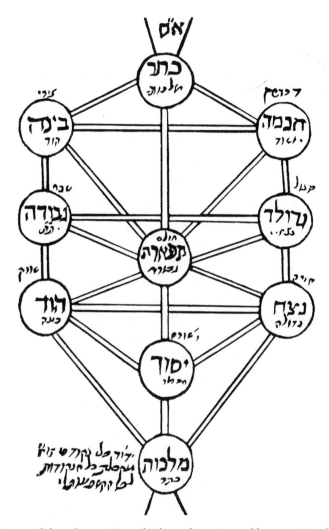

The tree of the *sefirot*, or Tree of Life, as shown in an old manuscript of the *Zohar*. The doctrine of the 10 *sefirot* influenced later ideas about the 4 worlds as developed by the Cabalists of Safed.

ascribe to him 10 virtues as a great ruler; cities and countries conquered by him appear in tens, and even the number of poets and jurists in his country was put at 10 or multiples of 10 so that the ideal number of perfection was reached.

More interesting, and spiritually important, than these well-intentioned games is the role of 10 in Ismaili gnostic writings. The Brethren of Purity had pointed out that 10 had a special importance as the first number with 2 digits (since they used the Arabic numerals and not, like the Europeans at that point in history, the clumsy Roman numbers). Classical Ismaili philosophy postulates 10 higher *hudud*, which can be described as intelligences or kinds of archangels. These 10 consist of 3: the primordial One, the first dyad, and the third intellect, plus 7 cherubim. From the first degree, that of *rasul*, "messenger," who, as the speaker, has to promulgate the divine Law and who is connected with the farthest heavenly sphere beyond the spheres, the descending line goes to the *wasi*, the "entrusted heir" (responsible for the esoteric interpretation) and the imam (connected with the sphere of Saturn) and continues downward until the sublunar world appears as the tenth step, connected with the lowest rank in the order of spiritual guides. After the Prophet and the entrusted heir, who in the historical manifestation is ʿAli ibn Abi Talib, follow the 7 imams to Muhammad ibn Ismaʿil. He will be followed by the *qaʾim az-zaman* as the tenth in line, and he is the one who will introduce resurrection. It is possible that Iranian gnostic traditions have survived, in an ingeniously islamicized form, in the Ismaili tradition.

For the rest of the Shiites, 10 has still another meaning. The Prophet's grandson Husayn ibnʿAli was killed by government troops in Karbala, Iraq, on 10 Muharram 61 (corresponding to 10 October 680). Already observed as a day of fasting under the name of ʿAshura during the Prophet's lifetime, this day is for millions of pious Shiites the great mo-

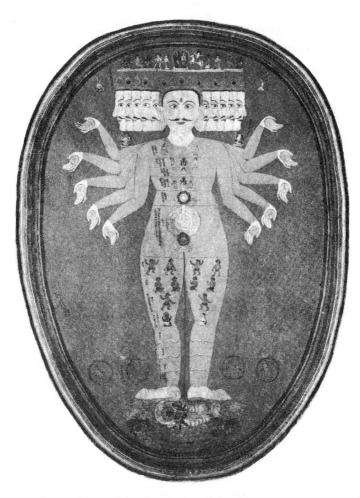

Tantric figure of the world with 10 arms and the 10 incarnations (avatars) of Vishnu. Gouache on cloth, Nepal, seventeenth century.

ment of suffering, comparable to the Christian Good Friday. During Muharram, *deh majlis*, "10 meetings," are often convened to recall the sad fate of the martyrs of Karbala, and numerous pious treatises about the sufferings of the imams are composed in 10 chapters or parts.

But we have to return to less mystical subjects. Ten is also known, and perhaps with even greater importance, for the decimal system, especially in connection with the military. The Turkish *on bashi*, "leader of 10," is followed by the *yüzbashi*, leader of 100, who corresponds exactly to the Roman *centurio*. In ancient Rome, the *decan* had to command 10 soldiers, although, of course, that is no longer the case when we use the terms *dekan*, *dean*, or *doyen*, which are derived from this designation. The word *decimate* reminds us of the custom that, when a Roman army revolted, every tenth soldier was executed. The *denarios*, *dinar*, points to the decimal counting of money. However, when a Persian poet speaks of a lily with 10 tongues, he intends to say metaphorically that this flower has a great number of petal-tongues, by which it expresses its silent praise of God. The Austrian orientalist Joseph von Hammer-Purgstall (d. 1856) was disappointed that no naturalist working in Iran could provide him with such a 10-petaled lily . . . But 10, like so many other larger numbers, serves frequently as a round number.

Finally, it may be mentioned at random that psychologists have interpreted 10 as 2×5, hence as the symbol of marriage, because 2 transforms eros, 5, into a more sober attitude.

Plate 1. The Creator measures the world according to size, number, and weight. The picture illustrates Proverbs 11:21. Augustine too taught that "All beings have shapes because they have numbers. If you take these numbers away from them they will be nothing. From where else are they than but from where is the number. For in them is as much of the essence as has been measured out" (*De libero arbitrio*, book 2). Manuscript illustration from a thirteenth-century French Bible.

Plate 2. Albrecht Dürer's famous engraving *Melencolia I* (1514) is filled with secret symbols. Its ambivalent, saturnine character reflects a turning point in Dürer's life. The magic square in the upper right is a so-called Jupiter square; it transforms into positive values whatever appears negative if seen under the saturnine aspect. The year when this engraving was executed was the one in which Dürer's mother died; the date of her death, 17.5.1514, is contained in the magic square.

Dürer's engravings *Rider, Death, and Devil* and *Jerome in his Cell*, which were executed at the same time, make us understand which problems plagued the artist. "It is the countenance of old Saturn that looks at us, but we have the right to recognize Dürer's own face in it as well" (Panofsky-Saxl).

Melencolia I was made for Emperor Maximilian, who was regarded as "Saturn-fearing"; it was meant to help him against melancholy, gloom, and sadness.

Plate 3. "Omnia nodes arcanes connexa": Everything is connected to everything by secret knots. Such was the motto of Athanasius Kircher, who, in 1665, composed a book called *Arithmologia sive de abditis numerorum mysteriis* in which he examined magic squares for the first time.

This plate shows the title page of an earlier book, *Magnes* (Rome, 1641). The circular symbols, grouped around the "eye of God" in the center, constitute a Tree of Life. Kircher's motto is written on the banderole above the eye of God.

Plate 4. "Omnia ab uno": all life comes from the One and through the One; Adam is created by God's breath. Mosaic from the cathedral of Monreale, Sicily, twelfth century.

Plate 5. The fourth trumpet, which announces the end of the world. Miniature from the Apocalypse of St. Sever (Gascony) according to the description in Revelation, chapter 8.

The 4 withered trees in the lower part of the manuscript allude to the situation on earth at that moment. When the seventh trumpet is blown, life will end completely.

Plate 6. Six cherubim with 6 wings each singing the "Holy holy holy" before the Lord, who sits on the throne above the Temple, a representation of Isaiah's vision (6:1–3). Miniature from a Torah manuscript in the treasury of the cathedral of Bamberg, *ca.* 1000.

Plate 7. The reconstitution of the Perfect 10: ascent to the kingdom of heaven. Miniature illuminating the "Song of Songs," School of Reichenau, *ca.* 980.

Wolfram von den Steinen interprets this image as follows: "In front of the Lord's throne, the orders of the heavenly hosts keep watch, each order in a symbolic group of three. They are nine altogether, but ten was the Golden Number of the Creator, which had been impaired by Lucifer's fall and should now be restored by the sanctified human beings. The vision in the picture shows how this will be achieved: on the lower level, the virgins come from the left, the women with their children from the right—again, three of each class. From the center, the men are ascending in pairs, laiety and priests, and a group of boys stands before them. All of them are led by Ecclesia, the Church, lavishly ornamented, beside whom a winged messenger appears."

Plate 8. The 24 Elders worshiping the Lamb (Revelation 4:4). Miniature from northern France, *ca.* 870.

THE MUTE NUMBER

○ ○ ○ ○ ○ ○ ○ **11**

Elf ist die Sünde. Elfe überschreiten die zehn Gebote.
(Eleven is the sin. Eleven transgresses the 10
Commandments.)

In placing this verdict in the wise astrologer Seni's mouth in
his drama *The Piccolomini,* Schiller voices a traditional opinion, for 11 has usually been connected with something negative. Larger than 10 and smaller than 12, it stands between 2
very important round numbers and therefore, while every
other number had at least one positive aspect, 11 was always
interpreted in medieval exegesis *ad malam partem,* in a purely negative sense. The sixteenth-century numerologist Petrus
Bungus went so far as to claim that 11 "has no connection
with divine things, no ladder reaching up to things above, nor
any merit." He considered it to be the number of sinners and
of penance. Medieval theological works often mention "the
11 heads of error." The Muslim Brethren of Purity also gave a
negative connotation to 11, regarding it as the first "mute"
number in the chain of "mute" prime numbers beyond 10.

Originally, 11 seems to have been connected with the
zodiac, for 1 of the 12 signs is always behind the sun, hence
invisible. This is alluded to in the Passover Haggadah (see p.
36), and in Joseph's dream (Gen. 37:9) where the sun, moon,

and 11 stars make obeisance to him, "stars" should probably be interpreted as "signs of the zodiac." The basis for such interpretation seems to be the ancient Babylonian creation myth as told in *Enuma elish*, which describes the struggle of Tiamat, the chaos, against the ordering gods, a struggle in which Tiamat is supported by 11 monstrous beings. These 12 adversaries are overcome by Marduk, the god of light; however, he does not kill them but rather, places them in the firmament and since he is the sun god, always stands before one of them.

In addition to these mythological associations, groups of 11 persons are also found in history, but it is next to impossible to explain the reason for their appearance. Among them are the Dionysiads, a group of 11 women in ancient Sparta formed to counteract the orgies of the Dionysian cult that were getting out of hand. In ancient Rome, a consortium of 11

Left: St. Ursula, protecting the 11 (thousand) virgins with her coat. Woodcut on the title page of *Dyt ist die wairhafftige ind gantze hystorie der hilligen XI dusent Jouffrauw enind mertelerschen* (That is the true and complete history of the 11,000 saintly virgins and their martyrdom) (Cologne, 1517). *Right:* Escutcheon of the city of Cologne, alluding to the virgins of St. Ursula with the symbol of 11 small flames.

men was charged with following criminals and detecting crimes—in other words, they constituted a kind of police force. But why does a soccer team have 11 players, the perfect 10 plus the goal keeper? And why do the Germans call the penalty kick in this game the *Elfmeter* (11 meter)? Only the historian of games and sports seems likely to explain this use of the 11, unless one wants to follow a psychologist like Friedjung, who sees the 11 players in this and other games as an allusion to human imperfection.

On the other hand, René Guénon has lately attempted to emphasize the positive values of 11 by explaining it as "the great number of the *hieros gamos*," the sacred marriage between macrocosm and microcosm. According to this novel interpretation, 11 consists of the combination of 5 (which is 2 + 3) and 6 (which is 2 × 3).

The fact that in the German Rhineland the carnival season begins on 11.11 at 11:11 A.M. should be ascribed to the amusing form of this date rather than to any mystery hidden behind 11. And when, according to legend, St. Ursula traveled to Cologne in a fleet of 11 boats, each of which carried 1000 virgins, this number probably was meant as an emphasized round number, meaning "even more than 10," since 11 can be used to "seal off" the preceding 10.

THE CLOSED
CIRCLE

○ ○ ○ ○ ○ ○ ○ **12**

5 and 7, the sacred numbers,
They rest in the 12.

This is again a quotation from Schiller, and as in the case of
11, he has made the astrologer Seni in *The Piccolomini* the
spokesman of ancient wisdom, for 12 has a very large circle of
meanings and activities. It can be seen first of all as the product of 3 × 4, which makes it a combination in which both the
spiritual and the material are contained, and second, as the
sum of the two meaningful numbers 5 + 7. The interest in
the 12 is likely to have developed out of the observation of the
zodiac. As was well known in ancient Babylonia, the moon
wanders through 12 stations, and so does the sun. There are
also 12 northern and 12 southern stars, which were known in
antiquity, and they may have influenced the idea of the 24
judges of the living and the dead, along with that of the 12
gates of heaven and 12 other gates leading to the netherworld
(where the Egyptian sun god Re spends the night). Many of
the ancient civilizations, especially in the Near East, were
duodecadic, built on 12, and they divided the year into 12
months. The most important exception, as Willi Hartner has
lucidly shown, was the high cultures of Latin America.

The 12 signs of the zodiac, which also appear in Etruscan culture, have probably influenced numerous myths, legends, and fairy tales in which groups of 12 deities, heroes or important personalities and time spans of 12 hours, days, or years frequently occur. Thus, one has tried to equate the tasks of Heracles with the sun's path through the zodiac, just as one tried to connect the myth of the Argonauts with the zodiac. Such equations were very fashionable around the turn of our century; one has to be very careful, however, about seeing astrological equations everywhere in myth and legend.

The 12 became an important round number in the ancient Near East and the Mediterranean world. Historically speaking, the 12 tribes of Israel were never exactly 12, and yet they formed a unit. The Old Testament has a good number of twelves: the 12 fountains of water in Elim (Num. 33:9), the 12 gems on the priestly robe of Aaron (Ex. 28:9–12), and the 12 stones that Joshua took from the Jordan to place in the Tabernacle (Josh. 4:5) are only a few examples. The fact that Christ chose 12 apostles was connected, by exegetes like Tertullian, with these groups of 12 in the Old Testament. In the Revelation of St. John, moreover, the heavenly Jerusalem has 12 gates, and 12 × 12 elect participate in the Adoration of the Lamb.

An important role is played by 12 in Christian allegoresis as well. If one regards Christ as the day, then the 12 hours are related to the 12 apostles, who were themselves prefigured by the 12 tribes of Israel, the 12 patriarchs, the 12 minor prophets, and even by the 12 shewbreads in the Tabernacle. According to Augustine, since 12 consists of 3 × 4, it is the Apostles' duty to propagate the faith in the Trinity in the 4 parts of the world. Similarly, an old English poem describes the phoenix, Christ's symbol, as rising with the sun, bathing 12 times, drinking water from the fountain 12 times, and flapping its wings at the beginning of every hour.

Revolving Time, with the 12 signs of the zodiac (inner circle) and the respective occupations during the 12 months (outer circle). Squatting in the center is Annus (Chronos), the old but constantly rejuvenating deity of the year, bearing the sun in his left hand and the moon in his right. Around the outer circle blow the 12 winds, and in the corners are the 4 seasons in their respective garments. Drawing from a liturgical book of the monastery of Zwiefalten, *ca.* 1140.

The 12 apostles listen to the Sermon on the Mount. Miniature painting from the Gospel of the German emperor Otto III. School of Reichenau, *ca.* 1000.

Twelve white pigeons hover around 3 concentric circles from which emerges a threefold monogram of Christ. The 8-pointed stars in the 4 corners symbolize God's supreme power. Mosaic from the baptismal chapel of Albenga, second half of the fifth century.

In various ancient cultures, the number 12 formed the basis for the incredibly large numbers found in their mythologies. Thus, the most important number of pre-Columbian chronology is the *baktun,* which consists of 144,000 days, and in Babylonia a multiple of 12 years appears with the 432,000 years of antediluvian time. Like many other large numbers, this figure can also be derived from 60, which is central to ancient Near Eastern number systems.

An interesting example of 12 as a counting unit is mentioned by Herodotus, who theorized that the Greeks founded 12 cities in Asia and refused to expand their number because they had been divided into 12 nations when living in the Peloponnesus, and thus 12 seemed to form the normative number of state and city.

In ancient China not only was the zodiac with its 12 signs well known, and the year of 12 months, but a decimal cycle was combined with this duodecimal cycle—exactly as in Babylonia to produce a sexagesimal system. One counted 12 hours of the day, and celebrated the completion of the circumambulation of Jupiter, which occurs every 12 years. One also believed that humans have 12 sections of intestines (and we do in fact have 12 ribs). The old high god in China, whose throne was located at the celestial pole, had, according to myth, 3 sons, 4 wives, and 12 chamberlains, and here, the connection with astral mythology seems clear.

In Germanic tradition, the relation between 12 and the zodiac was apparently less important. On the other hand, the 12 days or nights of the Germanic calendar that were intercalated to bring the lunar and the solar year together again live on in popular piety and superstition. They were often regarded as dangerous, or at least somewhat mysterious. Everyone knows the customs connected with the 12 nights between Christmas and Epiphany on 6 January: the dreams of each night predict something that will happen in the month corre-

"About the 12 fruits of life." The letters of the steps of the ladder and the leaves around the crucified Christ are cut in wood; the other texts are printed with type. Religious pamphlet, printed in Augsburg by Günther Zainer.

sponding to that night. One avoids washing laundry lest Wodan and his demonic horde carry it off or the accursed spirits that roam about in those dark nights endanger the home. The livestock receive special fodder, and the letters C + M + B, abbreviations for the names of the three Magi, Caspar, Melchior, and Balthasar, are written on doors and stables. Midnight likewise has a special character as the time when animals and spirits can talk, and the treasure hunter (as described in Goethe's poem *Der Schatzsucher*) sees the guiding light as the clock strikes 12 times. But once this uncanny hour has passed, everything returns to its normal shape, and all the lovely figures turn into pumpkins. A fine example of the literary importance of 12 is Edmund Spenser's *Faerie Queene*, where the virtues and their knights appear in groups of 12, as do the evil aspects of life, and everything is connected with 12 tales each. Just as the folk song recounts the dance around the maypole:

> Some dallied in the way
> and bound themselves by kisses twelve
> to meet next holiday.

In the Islamic context, the importance of 12 is less evident than in the Judeo-Christian tradition. It was again the work of the Brethren of Purity to indulge in the study of the 12 signs of the zodiac, following Babylonian and Greek astronomers and astrologers. They found that the 12 signs can be divided into 3 groups: there are not only 3 signs each for spring, summer, autumn, and winter, but also 3 fire signs (Aries, Leo, Sagittarius), 3 water signs (Cancer, Scorpio, Pisces), 3 air signs (Libra, Gemini, Aquarius), and 3 earth signs (Taurus, Virgo, Capricorn).

More interesting than these merely factual divisions is the development in the Shiite community. Here, the majority traced the sequence of the imams, the true leaders of the

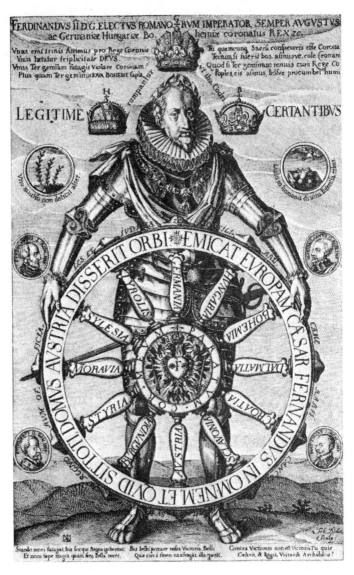

Ferdinand II, elected German emperor in Frankfurt in 1619, represented as the navigator of the realm. The spokes of the ship's wheel he holds represent the 12 provinces of his empire. Copperplate engraving by Tobias Bidenharter, 1620.

community, from the descendants of Prophet Muhammad, to the twelfth imam—hence their name, Twelver Shia (the group that has ruled in Iran since 1501). To what extent this line corresponds to a historical reality is difficult to decide, for the twelfth imam (whose very existence is doubted by some critical historians) disappeared mysteriously as a mere child in 874. For the Twelvers, however, he lives in hiding, guiding the world's affairs through his representatives until he will reappear to fill the world with justice as it is now filled with unjustice. It is therefore not surprising that Shiite scholars and poets give a special place to the number 12. The Persian historian and minister Rashiduddin, for example, executed in 1317, wrote his treatise on the great virtues of the 12 at the moment when his patron, the Mongol ruler Öljäitü, had just been converted to Twelver Shiism.

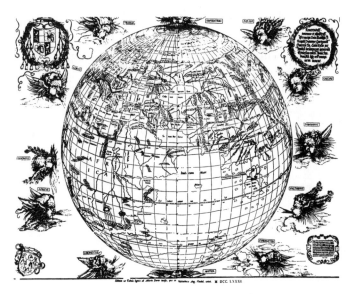

Twelve winds influence the globe. Map of the world by Albrecht Dürer, 1515.

To divide books, whether scholarly or literary, into 12 parts or 12 chapters is not rare among Muslim writers. The predilection for the 12 similarly led the Bektashi dervishes, one of the most important Turkish mystical fraternities and one known for its strong Shiite inclinations, to sport a head-dress with 12 wedges and to wear a duodecagonal agate, the Hajji Bektash stone, at their belt.

In Indian and Indo-Muslim tradition, one finds a special type of poetry, called *barahmasa*, "12 months," which has nothing to do with religious, much less Shiite, speculations but expresses the feelings and longing of a girl or young woman who is separated from her beloved or husband during the 12 months of the year; its use for religious purposes—the longing of the soul for the divine Beloved all year round—is a secondary development.

The duodecimal system has influenced our culture deeply, often in conjunction with the sexagesimal system: we still speak of the dozen, and we order (in Europe, at least) wine or mineral water in units of 12. In Germany one used to count the *Gros* (in English, the gross), that is 144 units, and the traditional English system of measures and weights was largely based on units of 12. The decimal system is slowly being introduced in the Anglo-Saxon world as well, but what to do, then, with a baker's dozen, which is 13?

LUCKY OR
UNLUCKY?

∘ ∘ ∘ ∘ ∘ ∘ ∘ **13**

In a fascinating article entitled "Triskaidekaphobia" (i.e., fear of 13), Paul Hoffman tells readers of the February 1987 *Smithsonian Magazine* that the phobia with this difficult name "costs America a billion dollars a year in absenteeism, train and plane cancellations and reduced commerce on the thirteenth of the month." Indeed, over the centuries, 13 has assumed an increasingly negative character in popular belief. One hesitates to invite 13 guests for dinner—even Napoleon, J. Paul Getty, and Franklin Delano Roosevelt were afraid of dining with 13 people at a table. (Apparently they did not know that this superstition can only be traced back to the seventeenth century.) Some hotels avoid the number 13 for rooms or even skip the thirteenth floor, just as there are railway stations in which a platform 13 is missing. And when the thirteenth day of a month happens to fall on a Friday (the day of Christ's death), it is viewed with even greater apprehension.

The Christian tradition explains this aversion to the 13 as a remembrance of the Last Supper, where one of the disciples—the thirteenth—betrayed Jesus. However, the negative role of 13 in Near Eastern civilizations and the cultures derived from them goes much farther back. Like 11, the 13 is a number that transgresses a closed system, in this case 12, the number of the zodiac: the sun never appears along with all 12 signs to make up 13 but rather, stands before one of them. This notion is reflected in fairy tales and myths where the hero is not supposed to open a thirteenth door, which would destroy the perfect, circular 12. In Babylonia the 13 had a certain negative aspect owing to its role in astronomy, and in China, likewise, it was connected with the division of the year. As long as a pure lunar year of 354 days was used, one had to intercalate a thirteenth month after a number of years to attune it again to the solar year; this month was called "Lord of distress," or "oppression."

In the Christian tradition, based on the 12 + 1 at the Last Supper, 13 was often mentioned as the number of the infernal hierarchies; likewise, witches frequently appear in groups of 13. It was also, as can be expected, connected more generally with witchcraft and black magic, and here the magic square relating to Mars plays an important role, as its central number is 13, and its sums always add up to 5×13. And when a German says, "Jetzt schlägt's Dreizehn" (Now, the clock is striking 13), it means that the closed circle of the 12 hours has been transgressed and "too much is too much!"

The German scholar Ernst Böklen has explored the so-called unlucky 13 and its mystical meaning in an exhaustive study (it is probably no accident that his book was published in Leipzig in 1913!). He shows that the groups of 13 in classical antiquity and the Middle Ages are mainly made up according to the pattern 12 + 1. (Gnostic theology similarly speaks of a thirteenth aeon that is supposed to bring the completion of the 12 previous ones.) As we mentioned in dealing with the 11, one could "fix" a number by adding 1 to it. This method can clearly be observed in the case of the Etruscan gods: the 6 divine couples are transformed into a unit by adding a thirteenth deity to them. Generally, the one beyond the 12 is either a leader or doomed to die. Thus, Philip of Macedonia, who had his own effigy paraded along with those of the 12 gods, was assassinated soon after this event, while Odysseus escaped death at the hands of the Cyclops but his 12 companions were devoured.

Groups of 12 + 1 are often found in European folktales, for instance, the type of 1 sister and 12 brothers (who are often transformed into animals so that the sister has to rescue them). In traditional Greek tales, Captain 13 is thrown into the abyss as the last member of the crew and hence remains alive. In France, the devil is said to snatch away every thirteenth person from a certain bridge before finally being overcome by the thirteenth member of a group of men. The type 12 + 1 often appears as a parent with 12 children, ranging from the biblical Jacob with his 12 sons to folk tales telling of the golden chicken with its 12 chicks or the gold-haired sow with her 12 piglets. Even the German law court maintains the old system of grouping 13 together in the form of the judge and 12 assistant jurors.

Admittedly, there is a certain ambiguity between 12 and 13, and in the role of 13 in general, in traditional tales, but in many cases 13 has an uncanny character. Death is the god-father of the thirteenth child, and a once-famous book by the German writer Friedrich Wilhelm Weber, which deals with a place called *Dreizehnlinden* ("13 linden trees"), is a deeply tragic tale filled with frightening, mysterious events.

While the 12 + 1 seems rightly to be traced back to Babylonian astral ideas, the number assumes a positive sense again in connection with the phases of the moon. In ancient Mexico, the lunar months were divided into 13 days of black moon, 13 days of full moon, and 13 days of new moon—in other words, the 2 periods of waxing and waning were counted at 13 days each. As was mentioned before, one imagined 13 heavens (above the 9 netherworlds and the earth) and, as a

According to the Maya calendar, the *tzolkin* or "counting of the days" consisted of 260 days. These were counted with 20 hieroglyphs which numbered 1 to 13. Thirteen marked the turning point, and its symbol was the butterfly. In the highest (thirteenth) sphere was situated the seat of the divine couple, "Lord and Lady of Duality" according to Aztec cosmology; their seat also signified a turning point. This picture shows the 13 deities of the hours, each of which is accompanied by a bird; on top, the butterfly.

corollary of this idea, 13 deities. Thus, 13 as the number of the heavenly spheres symbolized life, the sun, and the masculine power and was very positive and auspicious.

There are even more aspects of the 13 in ancient Maya religion. In one variant of the 19 numbers plus zero, called the "headed" variant, 13 numbers are distinguished by the sign of 13 different heads of deities. Furthermore, the 20 signs for the days of a month were combined with the numbers from 1 to 13, and thus a special calendar was devised for use in prognostication. Time was divided into periods of 52 (= 4 × 13) solar years of 365 days each; these in turn were summed up in 72 holy years, each of which consisted of 52 weeks with 5 days each (260 = 20 × 13).

As in ancient Maya culture, the 13 was also a sacred and auspicious number in the Hebrew tradition. The Cabala regards it as a lucky number since its numerical value in Hebrew (as in Arabic!) produces the word *Aḥad*, "One," the most important quality of God. From Exodus 34:6 one derived 13 divine qualities, and the Passover Haggadah (see p. 36) clearly emphasizes the importance of this number. An oracle in the Talmud claims that "at one time the land of Israel will be divided into 13 parts, the thirteenth of which will fall to the king Messiah." The Cabala speaks of the 13 heavenly fountains, 13 gates of mercy, and 13 rivers of balsam that the pious will find in paradise. The *Zohar* elaborates the importance of 13 when describing the three-headed Sacred Ancient of Days: "He can be found on 13 paths of kinds of love, because the wisdom hidden in him divides itself 3 times toward the 4 directions, and He, the Ancient One, comprises all of them." Here again we encounter the all-embracing principle that encloses the created world, which is determined by the 12 signs of the zodiac. One even went so far as to meditate upon the 13 conformations of the Holy Beard of the Divine, but a more practical and less esoteric occurrence of the 13 is that the Jewish boy celebrates his Bar Mitzvah at age 13 and is then under the obligation of the law.

It is somewhat surprising to see that medieval Christian theology has much less to say about the negative aspects of 13, outside the context of the 12 apostles and the "transgression" of this sacred number. More frequently, 13 was interpreted as a combination of 10 (the Commandments) and the 3 of the Trinity, or, with a different logic, as a combination of the Pentateuch (5) with the Resurrection of Christ (8). It thus was thought to point to the relation between the Old and the New Testaments, which was supposed to become manifest through a combination of work and faith.

While Europe increasingly disliked the number 13 after the Middle Ages, the French monarch Louis XIII was witty enough to declare it his favorite; he even married Anna of Austria when she was only 13 years old!

Thus, 13 offers many possibilities to the researcher, and perhaps its ancient positive aspects may help to cure some readers from the pains of triskaidekaphobia!

THE NUMBER OF
THE HELPERS

○ ○ ○ ○ ○ ○ ○ **14**

Higher numbers frequently assume a deeper meaning by being connected with their divisors. Regarding a number as a multiple of a lucky number, for example, gives the exegetes various ways to interpret it according to their purposes, as Meyer and other authors have shown in great detail. Some of the numbers have gained in importance by their connection with the lunar cycle; others are divisors or multiples of numbers derived from the 360 degrees of the circle.

Among the important lunar numbers, 14 is particularly interesting, for it takes the waxing moon 14 days to reach perfection as a full moon. In Babylonia one finds 14 deities who lead Nergal back into the netherworld, which may be derived from the days of the waning moon, but one can also interpret this 14 as a duplication of the 7 gates of the other-world. It may well be that a group of 14 helping saints, the *Nothelfer*, who are known in Catholic piety, belong to this tradition. But whatever their origin, devotion to them induced the pious to erect a beautiful baroque church, Vierzehn-heiligen ("14 saints"), in northern Franconia, as well as other chapels and churches. From here one may also look to the 14 innocent saints in Shiite Islam. The protective character of 14 innocent beings can be understood from an old German eve-

The 14 helpers or 14 saints are, in their aspect as doubled 7, symbols of kindness joined with reason, as well as symbols of help in dangerous situations. The monastery and church of *Vierzehnheiligen* (14 saints) in Franconia is the best-known example of their veneration. This woodcut appears on the title page of *Histori und ursprung der Wallfahrt und wunderzeichen zu viertzehn heyligen Nothelfern im Franckenthal* (History and origin of the pilgrimage and miraculous signs at the Fourteen saintly Helpers in the valley of Franconia) (1596).

ning song, set to music by Engelbert Humperdinck, which begins with the words:

Abends wenn ich schlafen geh,
vierzehn Englein mit mir stehn. . . .

(When at night I go to sleep, fourteen angels watch do keep. . . .)

Another use of the 14 comes from ancient Egypt, where the kindly god Osiris was, according to the myth, cut into 14 parts, and each of his parts brought a blessing to the land where it was lying.

It is natural that in Islam, a religion where lunar symbolism plays an important role, the 14 should occupy a special place. Once again the main contributions to the esoteric interpretation come from the Brethren of Purity. Fourteen is not only half of 28, the number of the lunar mansions and the letters of the Arabic alphabet, but it is also reflected in the 14 parts of the human hand and the 14 vertebrae each in the upper and lower parts of the spine. Is it not strange that the Arabic alphabet consists of 14 so-called sun letters and 14 moon letters, and also of 14 letters with diacritical marks and 14 without them? Such considerations played an important role in the mystical interpretations of the letters as it was elaborated in the late fourteenth century by the so-called Hurufis, who combined letter and number mysticism with the parts of the human face and body, showing, for example, that the words *yad*, "hand," and *wajh*, "face," both have the numerical value of 14.

It would be surprising if such speculations were not applied to the prophet Muhammad. Indeed, one of his names, Taha, has the numerical value of 14 and points to the fact that he, radiant like the full moon, appeared in the dark night of this world to illuminate it with his perfect spiritual and corporeal beauty. In this tradition, 14 is also connected with

beauty: 14 years is the ideal age of the beautiful young beloved, the boy with an immaculate face, comparable to a full
moon who, as a medieval Arab poet claims:

> is like the moon of 7 plus 7,
> and the 7 climata and the 7 spheres make obeisance before
> him. . . .

A LITTLE LUNAR NUMBER

○ ○ ○ ○ ○ ○ ○ 15

Fifteen represents the zenith of lunar power, and its relation to the moon can be deduced from the name of an old German measure called *Mandel*, "little moon, part of a moon," which consisted of 15 items such as eggs or other small things.

The 15 also has an important mathematical and religious meaning as the sum of the first 5 integers as well as the product of 2 sacred numbers, that is, 3 × 5. Fifteen was a number sacred to Ishtar, perhaps derived from the more important Ishtar-number, 5, perhaps also because it forms ¼ of the 60, the number of the highest sky god in Babylonia. Ancient Niniveh, the city devoted to Ishtar, had 15 gates, and the number of priests serving the Great Mother of Ida, which is again Ishtar, was 15. It is possible that 15 as the number of priests reached Roman tradition from Babylonia, for in ancient Rome the *quindecimviri*, 15 select men, were allowed to look at and interpret the sacred Sibylline Books. The fact that the Christian rosary is used to meditate on 15 mysteries connected with the life of Mary may well go back to the old number of female deities and saints, from Ishtar and Venus to Mary.

The Old Testament counts the generations of Israel between Abraham and Solomon as 15, and from Solomon to

The Ishtar Gate from Babylon in the time of Nebuchadnezzar II (604–562 B.C.E). Fourteen meters high, it is decorated with glazed relief tiles. These tiles contain 3 vertical rows of 5 sacred animals each, symbolizing Ishtar's sacred number, 15. (Staatliche Museen, Berlin)

Zedekiah again as 15. Here, the reflection of an old lunar myth cannot be excluded, since Solomon would then correspond to the full moon in all its glory and Zedekiah, who was blinded, to the dark moon.

Fifteen plays an important role in one of the most common magic squares which, built around the sacred 5, always offers 15 as a sum. Although legend attributes a Chinese origin to this square, it was known in Babylonia where it was connected with Ishtar. Combined with the star of Ishtar, with its 8 beams, the diagonals always add up to 15.

Symbol of
Wholeness

○ ○ ○ ○ ○ ○ ○ 16

Sixteen is a number of perfect measure and wholeness. In the Roman system 1 foot, *pes*, was formed by adding 4 palms, and each palm or breadth of the hand measured 4 fingers. Thus the foot was 16 fingers. A similar system is known from ancient Greece. In many languages there is a break after 16, as there is after 4: in Italian, after *sedicie*, 16, the form *dicia sette*, 17; in French, *seize* is followed by *dix-sept*.

Sixteen was a favorite number in India from time immemorial. It seems that it was used as early as the Indus civilization of the third and fourth millennia B.C.E., and until recently the rupee was divided into 16 anna. The Vedas mention sixteenfold incantations; as when one prepares the *soma*, the sacred intoxicating drink, and the *Chandogya Upanishad* claims that a complete human consists of 16 parts. One finds also 16 signs of beauty in classical Indian aesthetics, and the lovely lady is adorned with 16 different pieces of jewelry. Certain measures in poetry are divided into 16 *matra*, syllable units, and the most frequently used measure in Indian music is *tintal*, again with 16 units. It may be that the classical Arabic literature has taken over the 16 as a sign of perfection from Indian metrics, for some of the 16 classical Arabic

The 16-armed goddess Pussa, a Sino-Indian deity, sitting on a lotus flower. Copperplate from Athanasius Kircher, *China illustrata* (Amsterdam, 1667). Kircher regarded the deities of the Far East as variants of Egyptian and Babylonian gods and therefore considered Pussa a relative of Isis and Cybele.

"All of Nature consists of 16 elements according to philosophy." Diagram from *Geheime Figuren der Rosenkreuzer* (Altona, 1785–1788).

meters seem to serve merely to fill up the scheme in order to reach the perfect 16.

All those who were fond of the combinations and multiplications of the 4 elements and the 4 in general, as the number of orderly arrangement in time and space, have used 16 as the empowered 4 to express perfection—suffice it to mention the 4 × 4 philosophical elements of the Rosicrucians.

NUMBER OF CONQUEST

○ ○ ○ ○ ○ ○ ○ **17**

It seems unlikely that anyone in our day would be able to explain the deeper meaning of 17, and yet this number was quite important in the ancient Near East. The local god of Urartu, the area around Mount Ararat, received a seventeen-fold sacrifice, and perhaps 17 spread out from there into adjacent countries. In the Bible it is connected with the Flood, which began on the seventeenth day of the second month to end on the seventeenth day of the seventh month. At that point, Noah had reach Mount Ararat. One has speculated that 17 may be connected with the ark, or with floating on waters—in this connection one may think of Odysseus, who floated on a raft for 17 days after leaving Calypso. In Egypt, on the seventeenth day of a month the god Osiris is cast into the river in Typhon's coffin. There is also a Greek tradition that it is advisable to cut timber for boats on the seventeenth day of a month.

In classical antiquity 17 appears in connection with warfare and heroism. Naramsin, the Egyptian leader, invaded Syria and Asia Minor 17 times, and Thutmosis is credited with the same number of feats. However, it is a historical fact, though a strange coincidence, that Mahmud of Ghazna descended from the Afghan mountains into northwestern India 17 times between 999 and 1030.

Seventeen occupies an important place in the Islamic tradition. Centuries earlier, the Greek philosophers had been aware of the importance of the number: Aristotle, for example, discovered that the hexameter consists of 17 syllables, like the 2 central strings of the lyre with their 9:8 ratio, corresponding to the simple musical interval. Similarly, Poseidonius distinguished the 17 parts or faculties of the individual soul. Under the influence of this classical tradition, the first great Muslim alchemist, Jabir ibn Hayyan, saw the entire material world based upon 17: it consists, he noted, of the series 1:3:5:8, and these numbers, he claimed, form the foundation of all others. The traditional magic square around the 5 produces 17 in the lower left corner and 28 (the typical Islamic moon number) in the remaining numbers.

$$17 = \begin{array}{|c|c|c|} \hline 4 & 9 & 2 \\ \hline 3 & 5 & 7 \\ \hline 8 & 1 & 6 \\ \hline \end{array} = 28$$

The subsequent Islamic tradition, especially among the Shiites, discovered even more distinctive qualities of 17. The sum total of all the *rak'at* (the cycles of prayer movements in the 5 daily prayers) amounts to 17, and that is also the number of the words in the call to prayer. Some Sufis imagined that the greatest name of God consisted of 17 letters, and an early heretic, Mughira ibn Sa'id, who was executed in 737, claimed that at the appearance of the Mahdi, who will inaugurate the end of the world, 17 people will be resurrected first, each of whom receives 1 letter of the greatest name of God.

As Irène Mélikoff has shown, 17 appears very frequently in the tradition of the Turkish Bektashi order of dervishes.

'Ali ibn Abi Talib, the major religious figure in the order, had 17 companions and used to offer 17 prayers 3 times a day. There are also 17 saints who function as patron saints of the 17 Turkish artisans' guilds.

Popular Turkish legends, such as the tales of Abu Muslim or Malik Danishmand, mention 17 wars, as was the case in antiquity, and often 17 heroes are slain, or the leading hero receives 17 wounds. There has even been an attempt to prove that 17 is the sum of the digits of the numerical value of the names of most of the famous Turkish Muslim heroes—but that seems to be too far-fetched. Such speculations may well have been encouraged by another numerological circumstance, however: the initial letters of the names of Muhammad (m = 40), 'Ali (' = 70), and Salman (s = 50) add up to 170, the tenfold of 17. The trebled 17, that is 51, also appears in Shiite traditions; thus, for example, the treatises of the Brethren of Purity consist of 51 chapters. Seventeen is equally important for the Kurdish-Persian sect of the Ahl-i ḥaqq, and a medieval historian of Egypt mentions 17 as a lucky number.

In Christian tradition, the number appears, if at all, as a combination of the 10 Commandments with the 7 gifts of the Holy Spirit, described by St. Augustine as a *mirabile sacramentum*. Seventeen then forms the basis for the speculations about the 153 (see pp. 273–74). It also seems likely that the combination of *lex et gratia*, Law and Grace, as found in the 17 underlies the whole structure of the magnificent Golden Evangeliar made for King Heinrich III in the eleventh century, as Rathofer has shown.

THE DOUBLE NINE

○ ○ ○ ○ ○ ○ ○ **18**

The number 18, which appears in pre-Columbian measures of time, has a certain importance in the West as well. This may be explained by its connection with the circular number 360, or possibly with an astral number, since eclipses of the sun and moon recur in the same sequence after 18 years.

It is not surprising that 18 plays a role in the Cabala, for its numerical value corresponds to the word for *living* (*hay*), as well as the name David. In medieval Christian exegesis it is sometimes explained as the sum of 10 + 8, which is taken to mean the fulfillment of the Law through grace. This interpretation was derived from Luke 13:11, where it is told that Jesus healed a woman who had a spirit of infirmity for 18 years. Elsewhere, the number was taken as the product of 3 × 6, which represents faith in the Trinity coordinated with pious works. In some instances, 18 is used for arranging texts, as in the important Jewish prayer, the *Shmoneh-Esreh* (which means 18, referring to the 18 blessings of which it originally consisted) as well as the ancient Indian epic, the *Mahabharata*, which consists of 18 books.

Islamic tradition knows the 18 consonants of the introductory formula *Bismi'llāhi'r-raḥmāni'r-raḥīm* ("In the name of God the all-merciful, the all-compassionate"). Perhaps the common idea that there are 18,000 worlds is derived from the number of the blessed formula by which every act has to begin. For the Mevlevi, the Sufi order of the Whirling Dervishes, 18 is one of the central numbers and has a multiple meaning: the introductory poem of the *Mathnavi*, Jalaladdin Rumi's great didactic poem, consists of 18 verses, and everyone who wants to become a Mevlevi dervish has to serve for 18 days in the *tekke* (monastery) as a house boy and then learn the 18 different kinds of service in the kitchen. Once he has completed the 1001 days of preparation, he is led with an 18-armed candelabrum into his new cell, where he is supposed to devote himself to meditation for another 18 days. It was also the custom for visitors to a Mevlevi *tekke* to bring their gifts (such as candles) in numbers of 18. It is not clear whether this goes back to the traditional number of gifts among Turks, 9, but that is possible.

In the Germanic tradition, with its focus on 9, the duplication of the 9 appears as well. In Germanic mythology, Haldan had 18 sons, and Odin knew 18 things. One may also speculate whether the 18 great *arhat*, the leading saints of Chinese Buddhism, come from the duplication of a group of 9, or whether they are to be seen as 10 + 8, the combination of perfection and paradise. And even more difficult to explain is the fact that 18 little barrels of salt formed 1 load in medieval, predominantly English, tradition.

THE NUMBER OF
THE METONIC CYCLE

○ ○ ○ ○ ○ ○ ○ **19**

Nineteen is, to begin with, an incomplete 20, but is also a sacred number in the Near East. In ancient Egypt the *Book of the Dead* mentions the 19 limbs of the body, each of which had a deity of its own; by adding the deity connected with the whole body to them, the sacred 20 was attained. For medieval exegetes in the West, 19 combined the 12 signs of the zodiac with the 7 planets. In the Islamic tradition, however, this number corresponds to the numerical value of the word *wahid,* "One," which is one of the most important names of God. Recently someone in Pakistan has tried to prove that the Quran is based upon the number 19 (incidentally the number of guardians of hell) and its multiples. However, despite the clever use of computer technology for this study, its results have been challenged by most Muslims. On the other hand, 19 is uncontestably the sacred number of the Baha'is (again based on the numerical value of *wahid*), and they even divide the year into 19 months of 19 days each.

In ancient Babylonia, the nineteenth day of each month was considered to be unlucky, for this was the forty-ninth day from the beginning of the previous month, or 7 × 7 and that meant that it was filled with power, good or evil. Astronomically speaking, 19 is connected with the so-called Metonic cycle: every 19 years, all the phases of the waxing and waning moon fall on the same days of the week during the whole solar year.

An Old Limit of
Counting

○ ○ ○ ○ ○ ○ ○ **20**

To be sure, 20 is not exactly interesting for mystical and magic pursuits, but on the other hand, it is extremely important in the formation of number systems. For the simple reason that the fingers and toes add up to 20, this number forms the basis for counting in many cultures. Among the Celts, for example, 40 is called two times 20. The Ainu in northern Japan have a very similar way of counting and express, for instance, 80 as 4 times 20, which corresponds exactly to the French *quatre-vingts*.

The Maya used to connect 20 with the solar deity just as in ancient Babylonia it was linked with Shamash, the sun god. In Quiché, the Mayan writing system, the cipher for the number 20 is a human figure, and among the Hopi, a child is given a name on the twentieth day to become a real human.

In a number of special terms in German one still finds remnants of an earlier use of 20 as a comprehensive number: 1 *Ries* of paper, 20 sheets, forms a "book"; 1 *Stiege* consists of 20 eggs or sheep. The term *Stiege* can also be used for 20 yards of linen, and the *Schneise*, a small road into a forest, originally meant a rope on which 20 codfish could be hung to dry. Although few Germans remember these terms and their

The names and hieroglyphs of the 20 days. According to S. G. Morley, this is the oldest form of the Maya calender.

The Aztec calender, divided into 13 × 20 days. Each day has a name, such as crocodile (first day), wind (second day), house (third day), lizard, serpent, death, deer (seventh day), etc., with which the numbers 1 to 13 are connected. Besides this *tonalpohualli*, "counting of days," there was also a solar calender, *xihuitl*, whose 365 days were divided into 18 *cempohuallis*. The 5 *nemontemi*, "superfluous" days, were considered unfitting for any undertaking. The *cempohuallis*, of course, were not real "months" but artificial units connected with the most important number, 20. Being the sum of fingers and toes, 20 is "complete."

origins, in English, the old "score" for 20 is still much more in use, as in threescore and $1 = 61$.

At times, 20 appears as a round number, as in the *Odyssey*, when it is told that Odysseus cut 20 trees for his raft. In medieval Christian exegesis, on the other hand, 20 was taken to point to man's realization of the 10 Commandments in act and intention, or else as a product of 4 times 5, that is, the 5 books of the Pentateuch, which is the law, and the 4 Gospels, which mean grace. One may also find the explanation that it alludes to the four Gospels guiding the 5 senses.

PERFECTION

Twenty-one is connected with perfection, since it is the product of the sacred numbers 7 and 3.

THE HEBREW
ALPHABET

○ ○ ○ ○ ○ ○ **22**

In the cabalistic tradition, the 22 letters of the Hebrew alphabet are symbolized by the 22 almond blossoms on the 7-armed candelabrum of the Temple. They are classified as 3 mothers (*aleph, mim,* and *shin*), 7 duplicated letters, which correspond to the planetary spheres as well as to the metals, and finally 12 single letters related to the zodiac. Thus, the ancient sacred numbers are brought into a unified system. There are 22 ways that the 10 *sefirot* are mysteriously connected, and one has even tried to explain the existence of 22 letters by regarding them as a combination of the 10 *sefirot* with the 12 signs of the zodiac.

It seems that the 22 cards of the tarot and their *arcana,* or mysteries, have grown out of such speculations. The 22 pictures on the cards, placed in certain configurations and interpreted according to their numerical symbolism, are supposed to indicate certain basic human situations. Pictures on the oldest tarot cards, which were used in fourteenth-century Venice, can be connected with ancient Egyptian ideas and cabalistic mysticism. The tarock game, a favorite pastime in Bavaria, is derived from the old mystical tarot.

Christian exegetes have studied the 22, and the 22 sections

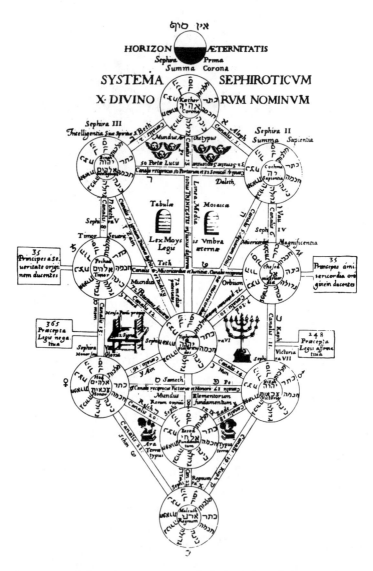

The system of the *sefirot* as the central glyph of the Cabala: 10 *sefirot* as manifestations of the Divine, 22 "true" ways (corresponding to the number of letters in the Hebrew alphabet), and altogether 32 paths of wisdom. Diagram from Athanasius Kircher, *Oedipus Aegyptiacus* (Rome, 1652).

The 22 major arcana of the tarot deck. These ancient cosmological playing-cards are used mainly to foretell the future. There are 21 consecutively numbered cards, plus the Fool, lower right, who represents 0.

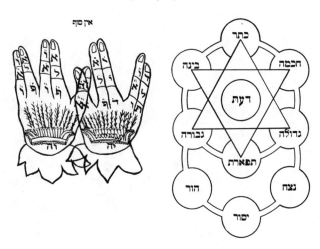

Left: Two hands on which the 32 connecting paths are marked. At the bottoms of the hands "YHWH," the name of God, is written. *Right:* Modern "Cabalistic tree."

of Augustine's *De civitate dei* point to 2 × 5 = 10 refutations, the negative precepts of the Law, and 3 × 4 positive teachings as manifested through the Trinity and the 4 Gospels and promulgated by the 12 apostles. Another Christian scholar, Sabas of Tales (d. 532), found that this number represents the number of God's works at creation, the books of the Old Testament, and the virtues of Christ. But, indeed, one should remember that the *Avesta,* the scripture of Zoroastrianism, contains 22 prayers as well.

TWENTY-FOUR TO
THIRTY-NINE

○ ○ ○ ○ ○ ○ ○ **24**

Twenty-four is the number of totality since it is connected
with the 24 hours of the day and night (although in antiquity
the day was counted in 12 double hours of 120 minutes each,
and in nautical reckoning, the glass, equaling 30 minutes measured by the hour glass, was maintained till recently). Twenty-four also appears as a measurement: the ancient German
Elle, or yard, consisted of 24 fingers' breadth. One can explain
the importance of 24 by interpreting it as the product of 2
pairs of integers: 4 × 6 and 3 × 8, and both strategies have
been much used in biblical exegesis.

Pythagoras regarded 24 as embracing the totality of the
parts of heaven, because of the 24 letters of the Greek alphabet as well as the 24 musical notes: by dint of the musical
sound, which can be explained by numerical relations, the
human being may be led to understand the harmony of the
spheres. In a similar vein, 24 is a most important number in
the Book of Revelation, where the 24 Elders embody the harmony of priest and king. The number can therefore be explained as representing the great harmony between heaven,
12, and earth, 2.

Twenty-four occurs in popular rhymes as well. In a fif-

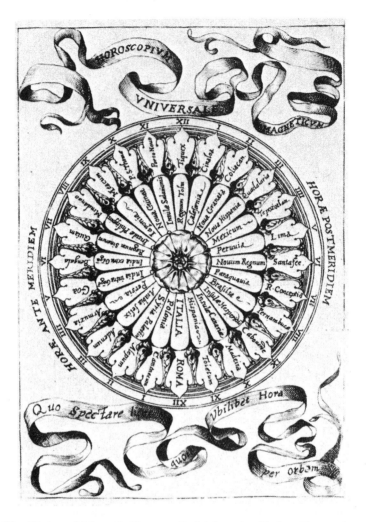

The "Universal Magnetic Horoscope" of Athanasius Kircher. Twenty-four
Jesuit houses and provinces are situated in the Far East and Latin America
but also in Spain and Poland; they are connected with the center of the
Society of Jesus in Rome. In fact these are the stations from which Kircher
obtained his information about distant countries and the customs of exotic
peoples. Copperplate from Athanasius Kircher, *Magnes sive de arte mag-
netica* (Rome, 1641).

teenth-century German verse about time, a tree with 12 branches bearing 30 nests each symbolizes the year; since the day has 24 hours, each nest contains 24 eggs, and while a black rat and a white rat are gnawing at the tree, time, in the form of a cat, finally devours everything. And of course there is the familiar English song:

> Sing a song o'sixpence,
> Pocket full of rye,
> Four and twenty blackbirds
> baked in a pie!

○ ○ ○ ○ ○ ○ ○ **25**

Twenty-five is the square of the sacred number 5 and is therefore used in the center of magic squares. It is also the sum total of 1, 3, 5, 7, and 9 and thus contains all the sacred numbers that can be used in magic. For this reason, in esoteric assemblies 25 candles would be lit so that every number necessary for mystical or magic practices was represented.

Medieval Christian exegetes tended to regard 25, the square of 5, as the perfection of the 5 senses. Alternately, they explained it as $6 \times (4 + 1)$, which meant: "good works on the basis of the 4 Gospels and the faith in one God." When presented as $(3 \times 8) + 1$, it was interpreted as the sign of the hope for resurrection, which is realized through faith in the 1 God who manifests himself as the Trinity. But there are also less mystical aspects of the 25. The quarter, 25 cents, is still a common coinage of the dollar.

Twenty-five is frequently used as a round number and serves, as a quarter of a full century, as a fitting number to celebrate silver jubilees.

○ ○ ○ ○ ○ ○ ○   **27**

Twenty-seven is interesting, from the mathematical view-point, as the third power of the sacred 3, thus, as Plutarch noted, as first odd, or masculine, cube. It also belongs to the lunar numbers, since the moon is at best visible 3 × 9 nights. Like 18, 27 appears frequently in traditions where 9 was of importance. Augustus prepared 3 × 9 sacrifices for the spirits of the netherworld, and in Russian folktales the hero often traverses 3 × 9 countries before reaching his goal. The number is common in the folk traditions of eastern Europe, such as collecting 27 flowers on St. John's night for protection or prognostication. A folksong from Lithuania describes the planting and growing of 3 × 9 rue, again for protective pur-poses.

Twenty-seven is mentioned in ancient Egypt, but in a negative sense. In a board game with 30 squares, the twenty-seventh, called water, means loss: the player who lands on it loses the game, and it has been speculated that dim memories of the dark moon are reflected in this negative result.

○ ○ ○ ○ ○ ○ ○   **28**

Twenty-eight is much more frequently used in connection with the moon because the 4 phases are completed once it has wandered through the 28 mansions. Besides, 28 is a perfect number from the arithmetical viewpoint because it can be seen as the sum of its divisors: 1, 2, 4, 7, and 14. This double significance, known since antiquity, was taken over into later traditions, culminating in the idea expressed by Albertus Magnus that the mystical body of Christ in the Eucharist appears in 28 phases.

Once again, 28 appears predominantly in religious traditions where 7 is central. A monument to Mithras in Siebenburgen, for example, repeats in 4 × 7 fields a dagger, a fire altar, a phrygian cap, and a cypress, thus totaling 28 objects connected with the Mithraic cult.

As a lunar number, 28 plays an important role in Islam, for mystics connect the 28 letters of the Arabic alphabet, in which the divine word, the Quran, is written, with the lunar mansions. The great medieval mathematician and historian al-Biruni (d. 1048) claims that this relationship proves the close connection between the cosmos and the word of God. It fits well into this picture that the Quran names 28 prophets before Muhammad, and poets therefore compare the Prophet of Islam to the full moon.

The numerical values of the 28 Arabic letters in the ancient Semitic order, the so-called *abjad* alphabet, can produce the whole series of numbers between 1 and 1000, for the first 10 letters form the one-digit numbers, 10 to 19 the tens, and the rest the hundreds, with the letter *ghayn,* the last one in this sequence, counting as 1000. Interestingly, the same way of producing 1000 out of 28 is found in the Egyptian *Thousand Songs of Thebes*, written around 1300 B.C.E., which consists not of 1000 but only of 28 poems.

Finally, it may be added that, as modern research has discovered, the epidermis is constantly regenerating itself, and all of its cells are replaced every 28 days.

○ ○ ○ ○ ○ ○ ○ **30**

Thirty is a number connected with order and justice. In ancient Rome, a man had to reach the age of 30 to become a tribune, and according to biblical traditions, both Moses and Jesus began their public preaching at that age. The Old Testament book of Judges ascribes to the judge Jair 30 sons who rode on 30 ass colts and had 30 cities (Judg. 10:4), while Samson promised the 30 friends at his feast that they would receive 30 robes and 30 festive garments if they could answer his riddle (Judg. 14:11–12). In the New Testament, meanwhile, Judas betrayed the lord Jesus for 30 pieces of silver. As a result some exegetes concluded that the value of any person, even a king, would be 29 *dirhams* or less, given that Christ's "price" was indeed not more than 30 pieces of silver. A delightful German ballad by Gottfried August Bürger (d. 1794) relates this story.

Another form of exegesis was used by early medieval Christian theologians such as the Venerable Bede to explain why, in the parable of the Sower (Mt. 13:8), the seed that fell on good soil brought forth grain "some a hundredfold, some sixty, some thirty." This was elucidated in terms of the old form of finger counting, in which the numbers 30 and 60 were expressed by certain figures made with the left hand, while 100 was a circle formed by the thumb and index finger of the right hand. 30 and 60, belonging to the left hand, are still incomplete, but 100 is fulfilment and eternal life.

○ ○ ○ ○ ○ ○ ○ **32**

Thirty-two, the fourfold 8, would appear to be a very happy and perfect number. Perhaps it was for this reason that the founder of the Hurufi sect in Islam, Fadl-Allah Astarabadi, claimed in the late fourteenth century to have perfected the Arabic alphabet with its 28 letters by adding 4 new ones (which are only used in Persian), thus reaching the number 32. The "happy" 8 also comes to mind in connection with the Buddha's 32 main and 80 secondary signs.

There are 32 cards in a pack of cards, and 32 pieces in chess. In Mongolian tradition, a folktale concerning Arji Borchi Khan describes 32 wooden figures standing on 32 steps, nicely arranged as on a game board. In the Cabala, 32 paths of wisdom emerge from the combination of the 10 *sefirot* with the 22 Hebrew letters; in the *Sefer Yetsirah* these are connected with the 32 teeth.

Thirty-two also appears among the measures; in Germany, for example, a *Malter* formerly consisted of 32 units, usually of ground material such as flour, as shown by its very name, derived from *mahlen,* to grind. The 32 points of the compass can easily be understood from the subdivision of the 4 cardinal points. But why did Fahrenheit place the freezing point at 32°?

○ ○ ○ ○ ○ ○ ○ **33**

Thirty-three is another number of completion and perfection. It is of special importance in the Christian tradition, since Jesus lived on earth for 33 years, just as David ruled for 33 years. The most conspicuous use of the mystical 33 occurs in Dante's *Divine Comedy*, which consists of 3 × 33 chapters. Thus the life of Jesus, symbolized by 33, leads toward God, yet falls one number short of absolute perfection (i.e., the One and, in Christian tradition, triune God).

Islam, meanwhile, knows the 99 Most Beautiful Names of God, which are often recited with the aid of a string of prayer beads usually made up of 33 beads.

In Indian mythology, 11 × 3 = 33 deities appears, again, as a perfect group, and multiples of 33 are also found in connection with divine beings. In the Celtic tradition, 33 seems to

The 33 steps of Freemasonry. The different ritual systems, including those of the women's lodges in the U.S., are represented. Lithograph by Everit Henry, late nineteenth century.

The Macedonian fresco in the monastery of Lesnovo (1349) shows Christ in heaven, surrounded by 33 likenesses that evoke the number of years of his earthly life.

be connected with black magic, perhaps as an inversion of the Christian sacred 33.

○ ○ ○ ○ ○ ○ ○   **35**

Thirty-five has no major mystical importance, but according to Plutarch it expresses harmony, as it is the sum of the first feminine and first masculine perfect cubes: 8 + 27.

○ ○ ○ ○ ○ ○ ○   **36**

Like 27, 36 has important arithmetical properties. Plutarch points out that it is a product of the first 2 perfect squares, 4 and 9, and at the same time, the sum of the first 3 cubic numbers, 1, 8, and 27. Furthermore, it can be regarded as a square number (6 × 6), but also as a rectangular one (4 × 9), and it is also the sum of the integers from 1 to 8.

Thirty-six is connected with astronomy, as it is a multiple of 12, and every zodiacal sign has 3 special aspects, known as the decans, which again total 36. In ancient Babylonia it was known that 36,000 years make a full circle of the polar star.

Nor should we forget that 36 is $\frac{1}{10}$ of the circumference of the circle.

That 36 has played an important role in China is understandable because of the 9 contained in it. It was considered to be a comprehensive, organizing number, and according to legend, the first Chinese empire consisted of 36 provinces and was surrounded by 36 foreign peoples.

○ ○ ○ ○ ○ ○ ○   **39**

Thirty-nine is an approximation of the round number 40. It plays an important role in the Jewish tradition because it permitted the avoidance of the number 40, which was often regarded as the utmost limits of acts or sins. Around the beginning of our era, a list of the 39 main kinds of tasks that must not be performed on the sabbath was prepared, for transgression against the sabbath would have demanded death by stoning. Later, 39 minor sins were added to each of these 39 main avoidable tasks. The real meaning was, of course, 40, the outside limit, but 39 gave a bit more freedom and lessened the danger of transgression. In fact, the comprehensive prohibition could be counted as the fortieth, all-embracing one. The same is true for punishments. As 40 blows were the upper limit of punishment, one permitted only 39 lest the limit be transgressed. A good example is found in St. Paul's letter to the Corinthians (2 Cor. 11:24) where the apostle says that he had received from the Jews "five times forty blows less one," meaning that he had been punished 5 times by beating.

PREPARATION AND COMPLETION

∘ ∘ ∘ ∘ ∘ ∘ ∘ **40**

Why did Ali Baba have to deal with 40 thieves? And why does Lent last for 40 days?

Among the higher numbers, 40 is by far the most fascinating one, being widely used throughout the Middle East and especially in the Persian and Turkish areas. From a purely scientific viewpoint, this number is associated with the disappearance of the Pleiades for 40 days, which was already observed in ancient Babylonia. This was also the length of the rainy season—as we are reminded by the 40 days' rain that caused Noah's flood. When the Pleiades returned from their "exile," the Babylonians celebrated a New Year's feast. Even today the weather is often predicted for 40 days according to the rule: "If it rains on this or that day we'll have rainy weather for the next 40 days." In old German farmers' traditions, this hold true for 8 June, 2 July, and 1 September, while the Day of the 7 Sleepers, 27 June, is taken to figure the weather for 7 weeks (i.e., 49 days).

One can also explain the importance of 40 by regarding it as a combination of the 28 lunar mansions with the 12 signs of the zodiac. The 40 large stone pillars in Stonehenge, arranged in a sacred circle with a diameter of 40 steps, seems to suggest an astronomical origin for the cult. (Incidentally, a combination of 28 kings or bishops with 12 other personalities to make up a group of 40, is quite frequent in the British-Germanic tradition.) Still another astronomical explanation of the importance of 40 can be found in the 40 aspects of Saturn which, according to the Bible, is the star of Judah. The number also has a biological role: pregnancy was formerly divided into periods of 7 × 40 days in order to observe certain changes in the embryo. Thus, according to the Islamic tradition the fetus is granted a soul after 3 × 40 days.

It seems that from the very beginning 40 has been a number connected with fate, and often with serious situations. The Old Testament claims the ideal length of human life to be 3 × 40 years (120 years), and many of the kings of Israel, including Solomon and David, were said to have ruled for 40 years. The time between the Exodus and the construction of the Temple was 12 generations of 40 years each (480 years). Some later Western scholars, such as Mahler, saw here traces of the "heavenly year" of ancient Near Eastern religions.

Medieval Christian exegesis found numerous allusions to the 40: from the 40 days of the deluge to the 40 years that the Children of Israel wandered in the desert, from the 40 days Moses spent on the mountain to the 40 days during which Christ was tempted by Satan in the desert. Similarly, Christ's period in the tomb lasted for 40 hours, which were explained by Honorius as pertaining to the revival of the 4 parts of the world: these had been dead as a result of the Decalogue and were now resurrected through Christ. The 40 hours that Christ rested in the tomb later give rise to the Roman Catholic "40 Hours Devotion," in which the Blessed Sacrament is exposed for a period of 40 hours, and the faithful take turns praying before it throughout this time.

Mathematically speaking, 40 is a *numerus abundans*, an

"abundant number" that can be divided by 1, 2, 4, 5, 8, 10, and 20, the sum of which (50) is larger than the original number. This fact gave rise to additional speculations around the combinations of the two numbers: the 40 days of Lent before Easter, for example, were taken to refer to the earthly life, while the 50 days between Easter and Pentecost point to eternal life, and so, virtuous life in this world leads to eternal rest and happiness.

St. Augustine interprets 40 as the product of 4, which points to time, and 10, which means "knowledge." Thus, 40 teaches us to live according to knowledge during our lifetime. One can also think of this mortal life in which humanity has to toil and labor in order to obtain the final consolation that Christ's appearance between resurrection and ascension—which lasted for 40 days—gives to believers. Again, it is possible to see 40 as the completion of the Law (the 10 Commandments) by the 4 Gospels.

More generally, 40 is the time of waiting and preparation, as becomes evident from biblical groups of 40 days or years. And to be the fortieth in a line is a hopeless situation, as John Donne has jokingly asked in "Love's Diet":

> What doth it avail
> to be the fortieth name in an entail?

Taking 40 as the completion of a stage of life, the Talmud, and later the Catholic church, declares it to be the "canonical age" of man, meaning that the intellect is then fully developed. Indeed, the modern psychologist often perceives a certain change in a person's development shortly before the onset of the forties: a look at biographies of famous people suffices to prove this point. In German, one calls this age *Schwabenalter,* alluding to the time when the inhabitants of the province of Swabia finally become mature. But the idea is not restricted to the West.

Along the same line one can mention Turkish expressions like "to play the lute after 40," which means starting something new and difficult at an advanced age, or "Stupid after 40, always stupid," which again points to 40 as the number of completion.

In Islamic lore, the importance of 40 is clear from both the Quran and the sayings of the Prophet Muhammad (who received his first revelation when he was about 40 years old). As in the Judeo-Christian tradition, the number is associated with the time of mourning or of patiently waiting. Popular mystical traditions claim, for example, that God kneaded Adam's clay for 40 days. When the end of the world approaches, the Mahdi will remain on earth for 40 years. At the resurrection, the skies will be covered with smoke for 40 days, and the resurrection itself is sometimes thought to last for 40 years.

In both Judaism and Islam, 40 days is the period of purification: after childbirth women remain confined for 40 days. In the Christian tradition, the feast of Candlemas on 2 February marks the end of Mary's confinement following the birth of Jesus and the completion of the required purification rites. Such rites are also considered necessary after the Islamic period of mourning, which lasts, again, 40 days. A modern development of purification may be seen in the *quarantine*, which originally lasted, as its name says, for 40 days. Purification plays another role in the Islamic tradition, where it is said that animals should be fed on special fodder 40 days before they are sacrificed; it is also recommended to cut one's hair and nails once every 40 days.

It goes without saying that such an important number could develop into a round number. Thus one finds groups of 40 throughout Muslim folklore: there are palaces with 40 columns (the garden pavilion Chihil Sutun, "40 pillars," in

Isfahan); heroes appear with 40 horses; mothers in fairy tales produce 40 children or 40 daughters in one birth. The hero has to go through 40 adventures or trials, kills 40 enemies, or finds 40 treasures. Frequently 40 martyrs are mentioned (this is also true for the Christian tradition, especially in Anatolia), and it is said that 40 brave men were slain at the Prophet's tomb in Medina. Muhammad's cousin and son-in-law, 'Ali, the first imam of Shiite Islam, had 40 disciples. In mystical Islam the 40 saints (*arba'in* in Arabic, *chihil* in Persian, *kïrk* in Turkish) play an important role; the Turkish town Kïrklareli, "country of the 40," still tells of its spiritual relations with such saints, and *kirklara karïşmak*, "to mix with the 40" means in Turkish "to become invisible" or to disappear completely. Forty is also an important round number for temporal events: the wedding feasts of heroes in Turkish or Persian folklore usually last 40 days and 40 nights.

Forty days or years often occur in proverbial sayings and popular customs. Thus, the medieval Arabic scholar Damiri claims that if a blue-eyed child were suckled for 40 days by an Abyssinian (black) wet nurse, his or her eyes would turn black. Likewise, a proverb current among Bedouins claims that someone who deals with the tribe's enemies for 40 days becomes one of them. If someone performs the morning prayer for 40 consecutive days under the lamp of the main mosque, he or she will be blessed by the vision of Khidr, the guardian saint of seekers after mystical enlightenment. In Sind, the southernmost province of Pakistan, a man who wants a woman to fall in love with him writes her name for 40 days on the leaves of a particular tree which he then throws in water, while an amulet for a new baby is secured by asking 40 men of the congregation in the mosque during the last Friday of Ramadan to write down the "Fatiha," the first sura of the Quran.

The popular Turkish character known as Nasrettin Hoca, whose jokes are told all over the country, has advised husbands to follow their wives' advice once every 40 years. In fact, "once every 40 years" means in Turkish "once in a lifetime," and when one drinks a cup of coffee with someone, the Turks claim, a 40-year relationship will be established. Can one change a die-hard sinner? No, for "you can put a dog's tail in a tube for 40 days and it still will not become straight." Another question heard in Turkey: "Is it not better to be a rooster for 1 day than to be a hen for 40 days?" Turkish folklore contains innumerable expressions in which 40 appears as the comprehensive round number. Our centipede (which, in German, has 1000 feet: *Tausendfüßler*) is known in Turkish as *Kïrkayak*, "with 40 feet," just as a wealthy person or a big landlord is "endowed with 40 keys." "The cat of 40 houses" is someone known everywhere while "the latch of 40 doors" is a jack-of-all-trades or else an impudent person who bothers everyone. To complete a long and difficult task one has to "eat bread from 40 ovens," and someone who sells himself all too cheaply, "turns 9 somersaults for 40 pennies." *Kïrklamak*, "to do something 40 times," means simply "to repeat frequently." Medieval Arabs believed that a person who went to the public bath (*hammam*) every Wednesday for 40 weeks would acquire all the riches of the world, and in modern times, unmarried girls in Baghdad still hope to find a husband by visiting the 3 main mosques of the city 40 times during the weekends.

In the Islamic tradition, 40 has another important function as well: it is the numerical value of the letter *mim* found at the beginning and middle of the Prophet Muhammad's name. Thus it is considered the typical number of the Prophet, all the more as it is also contained in his heavenly name, Aḥmad—and, as the Sufis discovered, when the *mim* is taken away from this name, the word *Aḥad* remains, and that

The letter *m*, which has the numerical value of 40 in the Arabic alphabet, is worked into the abstract design of this painting by the contemporary Pakistani artist Shemza.

means "One," an essential name of God. The difference between the divine One and the created prophet as humanity's representative was taken to point to the 40 steps that separate mortals from God and that have to be passed in the course of human development. These religious associations in turn induced Muslims to collect sayings and prophetic traditions in groups of 40: the *hadith* (sayings of the Prophet) were frequently put together in this way, and according to one of these *hadith*, the Prophet promises that, "Whosoever among my people learns by heart 40 *hadith* about religion will be resurrected at Doomsday along with the religious scholars and jurists." Such a group might include, for example, 40 sayings of the Prophet concerning a certain topic, such as divine mercy or the importance of writing, or 40 *hadith* transmitted by 40 people with the same name, or 40 that were collected in 40 different places, and so on. Such collections of "forties," as they are called (*arba'in*), were often artistically copied by master calligraphers.

A similar process was applied to the saying of the fourth caliph, 'Ali, and to verses from the great Persian mystical epic, Jalaladdin Rumi's *Mathnavi*. In mystical circles, the fortyfold repetition of religious formulas, particularly those concerning the name of the prophet Muhammad, was considered to be very effective. But this tendency to form groups of 40 is not restricted to religious literature: profane literature is also apt to include stories about 40 parrots or 40 viziers, or simply 40 stories together. Once again, this is mainly the case in the Persian- and Turkish-speaking areas. But throughout the Muslim world, the alms tax (*zakat*), which is one of the 5 Pillars of the religion, requires the contribution of $\frac{1}{40}$ of the believer's wealth to charity.

The old meaning of 40 as a number of preparation carries over prominently to Sufism. The comprehensive Arabic treatise *Ihya 'ulum ad-din* (The revival of the religious sciences) by the great medieval theologian and mystic Abu Hamid al-Ghazzali (d. 1111), for example, consists of 40 chapters leading the human being through the different stages of preparation up to the time, discussed in the final and fortieth chapter, when he meets his Lord at the moment of death. The Sufi is supposed to undergo a retreat of 40 days (*arba'in* in Arabic or *chilla* in Persian), a period of exclusive concentration on meditation and prayer. The Persian mystical poet Faridaddin 'Attar (d. 1220) has interpreted the experiences of the meditating mystic during these 40 days of seclusion in his epic *Musibat-nama* (The book of affliction). The pious repeated the *chilla* time and again, and it is a topos in Muslim hagiography to claim that a certain person had completed 40 *chilla*s at the time of his death. Even as I was working on this text, an American Jesuit wrote to me, apologizing for his long silence and explaining: "The last year was the most difficult one of my life. But what else could I expect? It was, after all, my fortieth, and He does not allow you to come out of a *chilla* and still think that you can do everything yourself—or?"

According to Augustine, 40 points to the *integritas saecularum*, the fullness of the times. And if one does not want to explain its importance in terms of its being a residue of ancient lunar myths, it can also be considered a "sanctified tetraktys" (as Paneth calls it): the sum of $(1 \times 4) + (2 \times 4) + (3 \times 4) + (4 \times 4)$, a number that contains the ideal Pythagorean measurement.

FORTY-TWO TO SIXTY-SIX

○ ○ ○ ○ ○ ○ ○ **42**

Forty-two is often interchangeable with 40, for 6 weeks of 7 days are often regarded, like the 40 days, as a time of preparation or waiting. Medieval Christian theologians pointed out that there were 42 stations that the Children of Israel had to pass between Egypt and Sinai, and that this sequence appears again in the incarnation of Christ in the forty-second generation after Abraham. Or else they might speak of the 6 connected with human acts multiplied by the 7 of resting, which is the sabbath, and see this as an allusion to the church in this world, living in hope of eternal peace. Forty-two is also the number of judgment, however, for the Egyptian *Book of the Dead* speaks of 42 judges who will examine the dead; it also enumerates 42 sins. In the Old Testament one reads that 42 boys were torn to pieces by bears because they had ridiculed the prophet Elisha (2 Kings 2:23–24), and Jehu kills 42 brothers of Ahaziah (2 Kings 9:27).

○ ○ ○ ○ ○ ○ ○ **46**

Forty-six is the number of years of the construction of the Temple (John 2:20) and also signifies the body of Christ, who

Ancient Egyptian representation of the Judgment of the Dead with the God of Death, the jackal-headed Anubis, and the Scribe-God, the ibis-headed Thoth, who notes down the deceased person's sins. The upper register depicts some of the 42 assessors of Osiris. There have been attempts to establish a connection between these 42 deities and the number of Egyptian provinces, but these only emerged in a later period.

appeared in this world $46 \times 6 = 276$ days after his conception. The numerical value of the Greek transcription of the name of Adam, which is $1 + 4 + 1 + 40$, led medieval exegetes to the same conclusion: 46 was connected with the human aspect of Jesus.

∘ ∘ ∘ ∘ ∘ ∘ ∘ **49**

Forty-nine, the second power of the sacred 7, marks the number of days after Christ's resurrection when the Holy Spirit was poured out upon the disciples. But since the beginning, or the end, of this period is counted as well, one speaks of the 50 days, *pentecoste*, and the feast of Pentecost. In Judaism, Shavuot is celebrated in a similar way on the fiftieth day after Passover to mark the giving of the Torah to Moses.

However, even before the introduction of these feasts, the number 49 was considered sacred since it pointed to God's rest (7 × 7 + 1).

50

Fifty, the enlarged 49, is associated with the jubilee year (Lev. 25:10), when feud and hatred were to cease in expression of the great sabbath of creation. But it has other meanings as well. Since the fiftieth psalm contains a complaint about its author's sins, for example, 50 is considered in allegorical exegesis as an expression of repentance and forgiveness: for the 10 Commandments are not always easy to obey for persons endowed with 5 senses (10 × 5 = 50). Fifty also plays an important role in the decimal system. It appears as an indefinite round number in both the Greek and the Irish traditions. When Hesiod speaks of 50 Nereids he means an enormous number of them, and the same is true for the 50 sons of Priam, the 50 cows of Helios, or the 50 pigs of Eumaios. In Irish folklore 3 × 50 frequently appears as a great round number, whether it be 3 × 50 queens, women, cows, pigs, or other creatures. King Arthur's table similarly consisted of 50 or 150 knights.

On the practical side, 50 was the age after which men were no longer required to perform military service in ancient Rome.

52

Fifty-two reminds us of the 52 weeks of the year, but its role was much more important in pre-Columbian America. The

Maya, in a complicated operation which they used to foretell the future, combined the images of the system of 13 with those of the vintagesimal system. The result was that every 260 days (52 × 5 or 13 × 20) the initial position of the numbers reappeared. Every 52 years, the beginning of the year corresponded exactly to the initial order. This was taken to mark a new beginning of the life cycle, and was celebrated as such. The importance of 52 is also reflected in an Aztec board game in which 52 fields are arranged in a cross shape. In medieval Europe, exegetes tried to explain 52 as a combination of the sacred numbers 40 and 12.

∘ ∘ ∘ ∘ ∘ ∘ ∘ **55**

Fifty-five is not considered a particularly important number today. However, it is interesting from an arithmetical viewpoint as the sum of all the integers between 1 and 10; it can also be divided into different elements, such as the sum of 28 + 12 + 10 + 5. For this reason it was used in some mystical speculations.

∘ ∘ ∘ ∘ ∘ ∘ ∘ **60**

Sixty, one of the central numbers in the systems of the ancient Near East, still influences our lives when we count minutes and seconds. It has also been used in other counting units, such as the old German measure called a *Schock*, which was mainly used for counting eggs but originally related to the placing of sheaves of grain at harvest time.

In Babylonia, 60 was the first great unit. As it could easily be divided by 30, 20, 15, 12, 10, 6, 5, 4, 3, and 2, it was, in a certain way, more practical than the 100 of the decimal system. While the number 1 was expressed by a small wedge, the 60 appeared as a larger wedge. Being the "great One" it was assigned to the highest deity, Anu, the god of heaven.

The relation of 60 to the 360 degrees of the circle and its use in the construction of geometrical figures such as the equilateral triangle may have played a role in the development of the sexagesimal system. Nor was its importance limited to ancient Babylonia: it was also considered a round number in ancient China, with 60 years as the normal lifespan of the individual. In ancient Greece, 60 was similarly used as a round number, and Plato's "marriage number," $12,960,000 = 60^4$, is derived from it. Sixty in the fourth power is the length of a world year (3,600 × 3,600 years), and every world day contains 36,000 years of human time. Plutarch claims that crocodiles lay 60 eggs, hatch them for 60 days, and live for 60 years; he adds that 60 is the first measure for those interested in heavenly appearances.

Sixty is also the product of 3 × 4 × 5, and the Kaaba in Mecca has recently been interpreted in these terms by the American musicologist and numerologist E. G. McClain, who relates its proportions to those of the Sumerian ark (60^3) or its double (2×60^3), the number of the sacred mountain, the world's center.

Talmudic speculations about 60 are more earthbound: fire is $1/60$ of hell, sleep $1/60$ of death, and the dream $1/60$ of prophecy.

○ ○ ○ ○ ○ ○ ○ **64**

If 8 is already a lucky number, connected with eternal beatitude and bliss, how much more auspicious its second power,

64, must be! Indeed, the number 64 is always related in some way or another to play and fate. According to ancient Indian doctrine, a man should master 64 different arts, and there are also 64 arts of loving taught in the *Kamasutra*. "From the eightfold combinations of the 8 kinds of embraces, kisses, scratches and bites, of couplings, etc., there emerge 8 groups of 8, that is 64," writes Richard Schmidt concerning the arts of love. But already in a much older Indian source, the *Rgveda*, the chapters are arranged in 64 sections. One also speaks of the 64 pleasures of Shiva, as if the 16, so important as the number of fullness and perfection, were reflected in a fourfold mirror.

Since 64 is the number of play and gambling par excellence, it is not surprising that the chessboard, which originated in India, should have 64 squares. Another important manifestation of the 64 is found in the Chinese *I Ching*, where the sixfold combinations of even and odd numbers result in 64 hexagrams symbolizing the variations of time and space. The German philosopher Leibniz discovered parallels between these 64 hexagrams and his own binary system, which, as

The 64 squares of the chess board can be seen as an arsenal of magic signs from alpha to omega, from A to Z.

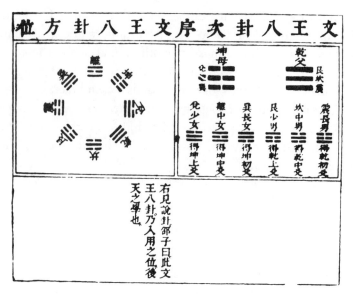

The *I Ching* consists of 64 hexagrams, created from every possible paired combination of 8 trigrams. Considered to be a description of the universe in abstract form, reflecting the perfect, closed system of categories in which the Chinese experience of the world is condensed, the *I Ching* for that reason has been and is still being used for prognostication.

Dyadik, was a forerunner of the binary system used today in computers. A contemporary scholar, Martin Schönberger, has even claimed that the millennia-old system of the *I Ching* anticipates the mathematical and philosophical knowledge used in the discovery of the first genetic code with its 64 triplets describing the DNA molecule.

○ ○ ○ ○ ○ ○   **66**

In Europe, 66 is known mainly as the name of a card game. However, in the Islamic mystical tradition it corresponds to

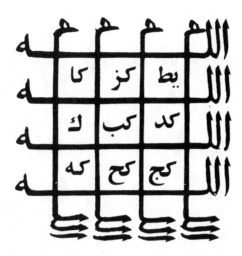

A magic square of Islamic origin that expresses the number 66 in every direction. The grid is formed by the letters of the word *Allah*, whose numerical value is also 66. Damascus, nineteenth century.

the numerical value of the word *Allah*. Hence, Turkish Sufis would explain the predominance of the tulip in Turkish art, as well as the emblem of Islam, the crescent moon, by the fact that the names of these 2 motifs, *lalah* and *hilal* respectively, consist of the same letters and have the same numerical value as *Allah*.

Sixty-six can also be seen as a duplication of 6, or as a number in the second power (as is the case with many similar numbers—44, 77, etc.). From a purely visual vantage point, 66 is 2 times 6, or 6 times 6. Interestingly, just such an interpretation appears in Turkish calligraphy, where the letter *waw* with its numerical value of 6 is often doubled. This motif is used in decorative inscriptions from the early eighteenth century on and can suggest either the number 66 (i.e., Allah) or $6 + 6 = 12$, which would be an allusion to the 12 imams of Shiite Islam.

The doubled Arabic letter *w,* which has the numerical value of 6. It may be interpreted, then, as 6 and 6 = 66, and thus pointing to the 66 of *Allah.* Alternately, it could be read as 6 + 6 = 12, which would then allude to the 12 imams of Shiite Islam. The *w* and the double *w* have been used in Turkey for calligrams since the early eighteenth century.

Seventy and Seventy-two: Plenitude

○ ○ ○ ○ ○ ○ ○ **70**

It is not surprising that 70 has a certain cabalistic value, since it is the tenfold of the sacred 7, and thus, as it were, its "great form." "Our life lasts threescore and ten. . ." (Ps. 90:10) says the psalmist, and groups of 70 are found throughout the Old Testament, from the 70 men accompanying Moses to Mount Sinai (Ex. 24:1) to the 70 years of the Babylonian exile (Jer. 25:12). This latter period also contains the aspect of mourning that is sometimes connected with 70—as when Moses was mourned for 70 days. Likewise, Isaiah threatened that Tyre would be forgotten for 70 years. According to Jewish and early Christian tradition, either 70 or 72 scholars are supposed to have translated the Old Testament into Greek in the days of the Egyptian king Ptolomy II, and the Greek version is therefore still known as *Septuaginta*, "of the 70." Whatever their real number was, since they were excellent scholars, they had to be a group of 70 or 72.

According to legend, Adam knew 70 languages, which is taken to mean all the languages of the world. In a similar vein, the Prophet Muhammad is said to have recited the Quran 70 times during his heavenly journey in the Divine Presence, and also to have asked forgiveness 70 times daily—a recurring motif in the Abrahamic religions.

A charming wordplay with 70 is found in the later *Midrash* where it is said: "He who remains sober while drinking wine has the insight of 70 wise men." The Hebrew word for wine, *yayin*, has the numerical value of 70, as does the word for secret, *sod;* thus, when the wine enters, the secrets come out.

Although 70 is itself a great round number, it can be multiplied. Islamic mysticism holds that 70,000 veils separate God from his creatures, and according to another mystical tradition in Islam, the Prophet's luminous essence stayed at the Tree of Knowledge for 70,000 years.

○ ○ ○ ○ ○ ○ ○ **72**

Much more important than 70, however, is 72. As ⅕ of the circle's circumference, it is related to the sacred 5, and because of its links with 5, 6, and 12, and also to 8, 72 became a favorite number in ancient times. Already in antiquity it was known that the vernal point of the sun advances by 1 degree of the zodiac every 72 years.

The medieval Cabalists taught that Yahweh's name consisted of 72 letters or that he had 72 names. The *Shem hamphorash* (the "unpronounceable name") is "the knotted numerological sum of all 72 mystical names of God." The 72 divine names were deduced from Exodus 14:19–21, where each of the 3 verses describing the extinction of the Egyptians who persecuted the Israelites consists of 72 letters: God has saved his people by means of his great name. In Christian exegesis, it was said that the 72 was the number of the bells on

Yahweh's 72 names in the form of a sunflower. According to the *Zohar*, "The crown of all legions rises in 72 lights." Diagram from Athanasius Kircher, *Oedipus Aegyptiacus* (Rome, 1652).

the priest's breastplate (mentioned in Exodus 28:17–21), which pointed to the 72 disciples of Christ who were called to preach the Gospel in the 72 languages of the world, each of which stands for one Divine name. One could, however, explain the number of these disciples in still another way: during the 24 hours of day and night they should not cease preaching the mystery of the Trinity in the world ($3 \times 24 = 72$).

As a number designating comprehensive multitudes, 72 occurs remarkably often in religious contexts. Chinese tradition knows 72 saints (8×9), and in Manicheism as in its late offspring, the Cathars, there were 72 bishops. Likewise, Islamic tradition claims that 72 people were martyred in the battle of Karbala, and in the religious tradition of the Yazidis, Adam is credited with 72 children (36 pairs of twins!) The best-known use of the 72 (or its augmented 73) in the Islamic tradition is the alleged existence of 72 or 73 Muslim sects, one of which is the "saved community." This concept is often alluded to in poetry, as when Hafiz states that he could not care less about the strife among the 72 sects as long as he has his glass of wine. Muslim legend claims that David was able to enchant the birds thanks to the 72 different sounds from his blessed throat.

Thus the number 72 appears everywhere, even in the tales of the *Arabian Nights,* to denote fullness composed of different elements, in contrast to the simple, undistinguished multitude expressed by numbers like 50. The use of 72 in this sense is not restricted to the Near Eastern world. It is equally important in China and Central Asia, for example: as a product of the 9 provinces and the 8 directions of the heaven, it is both auspicious and perfect. It was accordingly claimed that the wise Confucius lived for 72 years and had, like Jesus, 72 disciples to spread his teachings.

Under the dual inspiration of Islamic and Central Asian traditions, medieval Turkish authors expanded the number of

The 72 disciples of Kungfutse (Confucius), who lived for 72 years and according to legend had as many disciples as Christ (one for each divine name). Ancient Chinese stone rubbing.

heroes or adventures to 70 or 72. The Celtic-Germanic tradition mentions the latter number as well: the temple of the Holy Grail had 72 chapels (just as the big reception hall in Persepolis had 72 pillars). In Etruscan tradition, 72 years marked the real end of human life, and there was a rather widespread belief that the countries in the world numbered 72, by analogy with the 72 languages known—at least in legend—from antiquity.

In folktales, there is, for example, the Germanic story of *Wolfdietrich*, who was able to carry a gigantic woman over a distance of 72 miles. But it is part of history, and not fairy tales, that the historian Adam of Bremen found 72 sacrificial victims hanging from trees in Old Uppsala, Sweden after the end of a 9-day feast. Here, the importance of the 9, multiplied by 8, can be recognized very well.

In medieval German economy, meanwhile, 72 was seen as a combination of a *Schock*, threescore, plus a dozen.

UP TO TEN THOUSAND

○ ○ ○ ○ ○ ○ ○ **84**

Eighty-four is a favorite number in the Indian tradition. It is difficult to decide whether the number is considered a combination of the auspicious 8 with the 12, or whether, taking the visual aspects into consideration, it represents the auspicious 8 in the 4 corners of the world. The Nath Yogi's venerate 84 *siddhas,* or "accomplished ones," adepts who have attained immortality through their yogic accomplishments. More often, however, the number is multiplied by 1000. Thus King Ashoka is said to have built 84,000 stupas for the relics of the Buddha; the center of the universe, Mount Meru, is 84,000 units high, and many kings rule for 84,000 years. In the Jain system of thought, the basic number is 84,000,000.

○ ○ ○ ○ ○ ○ ○ **99**

Ninety-nine usually appears in contrast with the absolute unity, or else as something not yet perfect. Everyone knows the parable of the one lost sheep that was dearer to its owner than the 99 others. In Christian tradition, 99 also expresses

The "Muhammadan Rose." On the right flower, the 99 Most Beautiful Names of God; on the left, the 99 "noble names" of the prophet Muhammad. Turkish miniature, 1708.

the orders of the angels leading to God's unity. The same idea underlies the 99 Most Beautiful Names of God in Islam, all of which point to the comprehensive unity inherent in the all-embracing Greatest Name. Later Islam also developed the idea that there are 99 names of the Prophet. The ancient Christian church read the word *Amen* (1 + 40 + 8 + 50) as 99, but in Islamic tradition, where the *e* is replaced with an *i*, it would count as 101.

Ninety-nine can be used as a circumscribing number. The poets ask for 99 kisses plus one (as is the case with Imru'l-Qays in sixth-century Arabia), and the ancient Arab hero Shanfara vows to kill 100 enemies but is slain by the ninety-ninth; however, the splinters from one of Shanfara's bones injure this one, and he too dies, thus completing the number 100.

○ ○ ○ ○ ○ ○ ○ **100**

One hundred is the great round number of perfection. As the second power of the sacred 10 it designated the perfect Good in the Hellenistic world. In the decimal system, "100 times" usually means "often": "Haven't I told you 100 times not to do that?" The Chinese expression "100 mouths" is very fitting for the whole extended family one has to feed. More recently from China comes Chairman Mao's dictum, "Let 100 flowers bloom," which ushered in the Cultural Revolution. In terms of measure, there is the *Zentner* with its 100 (German) pounds (= 50 kg); this weight, as the *centenarium*, goes back to the old division of measures and weights according to the decimal system. Nevertheless, the exact number of pounds is not so important, for the *Zentner* could have different weights up to the "Great 100," which is 120. Nor did the *centurios*, the Roman captain, always have to command exactly 100 soldiers, just as the Mamluk *amir mi'a*, "leader of 100," did not always possess exactly 100 military slaves.

○ ○ ○ ○ ○ ○ ○ **101**

After the high round number of 100, 101 assumes special importance. Like 41 on the lower scale and 1001 in the next degree, it is a number of infinity. Religious formulas such as the blessings for the Prophet Muhammad are repeated 101 times, and in Indo-Pakistan it was customary to give a bride 101 pieces of clothing and 101 trays filled with gifts.

○ ○ ○ ○ ○ ○ ○ **108**

One hundred and eight, the product of 12 × 9, appears often in Hindu and Buddhist traditions. It may be derived from the frequently used 18. The *gopis,* or cowgirls with whom the god Krishna dances and plays, number 108, and, in enlarged form, one speaks of 108,000 degrees of reincarnation. For the Buddhists, 108 is the number of the *arhat,* the perfected saints, and prayer beads are also arranged in groups of 108. Likewise the Tibetan sacred scriptures, the *Tanjur* and the *Kanjur,* contain 108 parts. When a Tibetan Buddhist prince or princess died, no fewer than 108 lamas were supposed to be present at the funeral.

∘ ∘ ∘ ∘ ∘ ∘ ∘ **120**

In European tradition, 120 is called the Great 100, numbering 10 dozens. It also holds an important place in the Old Testament (Gen. 6:3), where it is connected with the lifespan, and even today one still hears the Jewish expression at the mention of someone's name: "So and so—may he [or she] live to be 120!"

∘ ∘ ∘ ∘ ∘ ∘ ∘ **144**

One hundred forty-four is the *Gros* or "fat dozen," that is, the second power of the sacred 12. Hence it appears wherever 12 should be multiplied. It has an important place in the Revelation of John (Rev. 14:1–5), where the Lamb appears on Mount Zion, surrounded by the 144,000 who have been redeemed.

∘ ∘ ∘ ∘ ∘ ∘ ∘ **153**

This appears to be a very curious number. Why, for example, did the disciples of Christ catch exactly 153 fishes, as John (21:11) reports? There is, first of all, a mathematical interest, since 153 is the triangular figure of 17, that is, it can be expressed as $1^3 \times 3^3 \times 5^3 = 1 + 27 + 125$.

In attempting to come up with an explanation for the number of fishes, medieval Christian theologians referred back to 17, which, as we saw earlier, contains the Law (10) and Grace (7). Again, interpreted as $3 \times 3 \times 17$ and as the sum of all integers between 1 and 17, 153 could indeed become a fitting symbol for the relation between Law and Grace, or between the Trinity, the Law, and the 7 gifts of the Spirit. According to Augustine, meanwhile, 153 represents the multitude of the faithful from all over the world who are captured by the fishers of men for the kingdom of God.

○ ○ ○ ○ ○ ○ ○ **216**

The number 216 is worthy of mention as that of auspicious signs on the Buddha's footprint. As the product of 8 and 27, it contains much beneficial power.

○ ○ ○ ○ ○ ○ ○ **240**

On the other hand, 240 has only a practical meaning and nothing mystical to it. Charlemagne minted 1 pound of silver into 240 *denarii,* and until recently, the English pound was divided in 240 units (20 shillings of 12 pence each).

○ ○ ○ ○ ○ ○ ○ **248**

This number is given a prominent place in Judaism as it represents the numerical value of the first two words of the fundamental religious expression of monotheism, *shmac Israel,* "Hear, O Israel." By extension, it corresponds to the 248 limbs of the human body, each of which should be permeated by one letter of the profession of God's unity.

○ ○ ○ ○ ○ ○ ○ **270**

In the *Zohar*, 270 stands for this world or for the Last Judgment, since it corresponds to the numerical value of *ra^c*, "evil, bad."

○ ○ ○ ○ ○ ○ ○ **300**

As the tenfold of the ordering 30, 300 appears in biblical tradition as a number connected with heroes. Abishai was the chief of 30 and wielded his spear against 300 men and slew them (2 Sam. 23:18); Gideon is also connected with 300 heroes, who were chosen by the Lord out of all Gideon's army to defeat the forces of the Midianites (Jg. 7:7). It can be used as a great round number as in the *Quinzevingts*, 15 × 20, which is the name of the hospital founded by Louis IX in Paris in the mid-thirteenth century for 300 blind people. In this case, the ancient French vintagesimal system is clearly visible.

○ ○ ○ ○ ○ ○ ○ **318**

This is mentioned in the Bible as the number of Abraham's servants. Since the only one mentioned by name is Eliezer, and his name has a numerical value of 318, this association can be easily explained.

○ ○ ○ ○ ○ ○ ○ **360**

The number 360 is highly important in relation to the degrees in the circle and the round year. For this reason one finds, to give only one typical example, 360 stones must be used to make an altar in a Vedic temple ritual.

The lunar number 27 and the solar number 360 have the ratio 1:13½, which, with small variations, is also the ratio between the value of the solar metal, gold, and that of the lunar metal, silver.

○ ○ ○ ○ ○ ○ ○ **432**

The number 432 breaks down into 4 × 108. Multiplied by 1000, it appears as an enormous span of time in Indian tradition; thus, 10 primordial kings ruled 432,000 years before the deluge. The cosmic cycle that the Hindus call the *mahayuga*, meanwhile, consists of 4,320,000 years, a number that is obtained by multiplying 4320 first by 4, then by 3, then by 2, and finally by 1; this sequence mirrors the steadily decreasing length of mythical eras (the first is 4000 years, the next 3000, and so on), and contains the tetraktys (4 + 3 + 2 + 1 = 10).

○ ○ ○ ○ ○ ○ ○ **666**

This is a biblical number that has nourished the imagination of generations of Christians and is still much discussed today. It occurs in Revelation 13:8 as the number of the Beast, and with some skill one can find it in almost every name by counting the numerical values of the letters (and often shifting them until the result is achieved.) One can interpret it as signifying a Roman emperor, a modern dictator, or whoever the *bête noire* of the moment happens to be. The interpretations of two leading numerologists of the sixteenth century, Michael Stifel and Petrus Bungus, are very typical. Stifel, a Protestant, saw the number pointing to Pope Leo X while Bungus, a Catholic, considered it an allusion to Luther. Among the other readings Bungus offered was the name of the Prophet Muhammad, *Mahometus*, which was not surprising in the days of Turkish conquests in Europe.

∘ ∘ ∘ ∘ ∘ ∘ ○ **1000**

In the decimal system, 1000 is, predictably, the all-embracing number. "A thousand years are before you like a day . . . ," sings the psalmist, and the believer is consoled, "Thousands may fall at thy right and tens of thousands at thy left. . . ." When the Chinese offer birthday greetings by wishing someone 1000 springs, it corresponds exactly to our "Many happy returns." One thousand turns up in German names of animals or plants such as *Tausendfüßler*, centipede, and *Tausendschönchen*, whose English name, daisy, sounds less romantic. The *Tausendsassa*, "jack-of-all-trades," belongs to the same group.

∘ ∘ ∘ ∘ ∘ ∘ ○ **1001**

Like 41 and 101, 1001 exceeds the great number and therefore comes to mean "infinite, numberless." One may think of

Left: Title page of the 12-volume German translation of the *Alf layla wa-layla* (1001 Nights), the best-known collection of Oriental fairy tales (Leipzig, 1907–1908). *Right:* Title page of the German translation of *Hazar u yak ruz* (1001 Days), a Persian collection completely different from the *1001 Nights*. The identical drawings by Marcus Behner do not reflect the different contents of the books.

Binbir kilise, 1001 churches, in Anatolia, or the best-known example, the *1001 Nights*. This number leads to infinite extension, although attempts have been made to explain it as an allusion to the heptagon and mysterious relations between the prime numbers 7, 11, and 13.

∘ ∘ ∘ ∘ ∘ ∘ ∘ **10,000**

Ten thousand is the upper limit in the decimal series. In China, 10,000 years means immortality, and the 10,000 things are everything that exists. When the Germans speak of

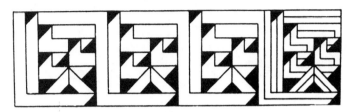

The glyph *10,000* as a stylized, meandering brook. According to Lao-tzu: "Out of 1 come 2, out of 2 come 3, and out of 3 the 10,000 things." Hence this number symbolizes the infinite multitude of manifestations in the universe and also immortality.

the *Obere Zehntausend* (the upper 10,000) however, they are referring to an exactly defined social elite.

In India, as among the Maya and Aztecs of Central America, higher numbers are frequently used and incorporated into speech. The length of Indian time spans in myths and tales is well known, and Indian Buddhists go even further in speaking of quantities far beyond imagination. Lately there have been attempts to establish relations between these enormous numbers and those used in modern physics, but nothing concrete has been established.

We cannot lose ourselves in such speculations, nor can we offer any recipe for successful number magic. Humans have tried to solve the mystery of numbers for millennia, using and misusing them, and yet, the fascination remains. As the architect Le Corbusier tells us: "Behind the wall, the gods play. They play with numbers, of which the Universe is made up."

BIBLIOGRAPHY

GENERAL WORKS ABOUT NUMEROLOGY AND NUMBER MAGIC

Ahmad, Mabel L. *Christian Names and their Values.* London, 1930. An attempt at explaining the numerological propensities of Christian names.

Allendy, R. F. *Le symbolisme des nombres.* Paris, 1921. A knowledgeable general introduction to the problems of numerology.

Ballieth, L. Dow. *Nature's Symphony, or Lessons in Number Vibration.* Atlantic City, 1911. Attempts to find correspondences between letters, numbers, sounds, colors, and so on.

Bell, E. T. *Numerology.* New York-London 1933, 1946. The American mathematician sharply attacks numerological magic as it was spreading in the United States in the early 1930s.

Bindel, Ernst. *Die geistigen Grundlagen der Zahlen.* Stuttgart, 1958, 1975.

Bischoff, Erich. *Die Mystik und Magie der Zahlen.* Berlin, 1920. A brief introduction to and lesson in how to produce magic squares, along with "a systematic symbolism of numbers from 1 to 4,320,000."

Blankenhagen, Walter, and Willem Enders. "Zahlensymbolik." In *Musik in Geschichte und Gegenwart (MGG),* Supplement, Kassel, 1979, cols. 1971–1978. A fascinating survey of numerical symbolism in music, with a rich bibliography.

Bressendorf, Oskar von. *Zahl und Kosmos.* Augsburg, 1930.

Brinton, D. G. "The Origin of Sacred Numbers." *American Anthropologist* 7 (1894), pp. 168–173.

Bungus, Petrus. *Numerorum Mysteria.* Bergamo 1599; reprint, with a very useful introduction by Ulrich Ernst. Hildesheim, 1982.

Castro, Boanerges B. *O simbolismo dos numeros ne Maçonaria.* Rio de Janeiro, 1977. About number symbolism among the freemasons.

Cheiro. *Das Buch der Zahlen,* 6th ed. Freiburg, 1981. German translation of the "Book of Numbers," a popular book about occult meanings of numbers.

Chevalier, Jean, and Alain Gheerbrant. *Dictionnnaire des symboles.* Paris, 1969.

Clichtovaeus, Jodocus. *De mystica numerorum significatione.* Paris, 1513. Reprint. Hildesheim, 1985.

Crawley, E. S. "The Origin and Development of Number Symbolism." *Popular Science Monthly* 51 (1897), pp. 524–534.

Dornseiff, Franz. *Das Alphabet in Mystik und Magie.* 2d. ed. Leipzig, 1925. The classic study of letter magic and related fields.

Ellis, Keith. *Numberpower: in nature, art, and everyday life.* London, 1977.

Emrich, Louis. *Magie der Zahlen; der Mensch und seine Glücks- und Unglückszahlen.* Büdingen, 1952.

Feldmann, F. "Numerorum Mysteria." *Archiv für Musikwissenschaft* 14 (1957), pp. 102–127.

Forstner, Dorothea, OSB. *Die Welt der Symbole.* 2d ed. Innsbruck, 1967. Introduction to Christian symbolism by a Benedictine nun.

Friedjung, Walther. *Vom Symbolgehalt der Zahl.* Vienna, 1968. A somewhat confused book in which one finds interesting and absurd remarks side by side.

Garcia Gomez, Emilio. "La magia de los numeros." ABC (Madrid), 25 June 1985.

Gardner, Martin. *The Numerology of Dr. Matrix: The Fabulous Feasts and Adventures in Number Theory, Sleigh of Word, and Numerological Analysis.* New York, 1967.

———. *The Incredible Dr. Matrix.* New York, c. 1976.

Ghyka, Matila C. *Philosophie et mystique du nombre.* Paris, 1952.

Hara, A. H. *Number, Name and Colour.* London, 1907.

Heiler, Friedrich. *Erscheinungsformen und Wesen der Religion.* Stuttgart, 1961. The chapter on "the sacred number" ("Die heilige Zahl"), pp. 161–176, in this comprehensive phenomenology of religion is extremely rich, with important bibliographical notes.

Hellenbach, Lazar Baron von. *Die Magie der Zahlen als Grundlage aller Mannigfaltigkeit.* Leipzig, 1910. 2d. ed. 1923. Contains all kinds of true and false information.

Kircher, Athanasius. *Arithmologia, sive, de abditis numerorum mysteriis.* Rome, 1665. Reprint. Hildesheim, 1984.

Kissener, Hermann. *Lebenszahl: die Logik von Buchstabe, Zahl und Zeit.* Munich, 1960, 1974. An attempt to combine different numerological systems.

Kükelhaus, Hugo. *Urzahl und Gebärde. Grundzüge eines kommenden Maßbewußtseins.* Berlin, 1934. Reprint. Zug, 1980.

Lawlor, Robert. *Sacred Geometry.* London, 1982.

Lüttich, Selmar. *Über bedeutungsvolle Zahlen.* Jahresbericht Domgymnasium. Naumburg, 1891. A well-researched and quite interesting article with many useful examples.

Marx, Otto. *Die Zahlen in der Volksheilkunde des schlesischen Raumes.* Breslau, 1934. A welcome analysis of the practical application of number magic in the province of Silesia (now Poland).

McClain, Ernest G. *The Myth of Invariance. The Origin of the Gods, Mathematics and Music from the RgVeda to Plato.* New York, 1976.

———. *Meditations through the Quran.* Tonal Images in an Oral Culture. York Beach, 1981. Both works of the musicologist McClain contain very fascinating ideas, but the author seems to lose himself in speculations that are difficult for a lay reader to comprehend.

Nicolas, A. T. de. *Meditations through the Rg Veda. Four-Dimensional Man.* Stony Brook, 1978. Interesting, though highly speculative.

"Numbers" in *Encyclopedia of Religion and Ethics,* 1911, vol. 9, pp. 406–417. Three separate entries including an overview, the Arian use of numbers, and the Semitic use of numbers.

"Number Games and Other Mathematical Recreations," in *Encyclopedia Britannica,* 1974, vol. 13, pp. 345–357. An instructive and amusing article.

Paneth, Ludwig. *Zahlensymbolik im Unbewußten.* Zurich, 1952. Study by a Jungian psychologist about numbers in dreams and the character of numbers; counters Freud's analysis of dreams.

Riefstahl, Hermann. "Ontische Studien." *Beiträge der Humboldt- Gesellschaft* 4 (1983), pp. 155 ff. Explanation of the first numbers from a philosophical vantage point.

Riemschneider, Margarete. *Das Geheimnis der numinosen Zahl.* Munich 1966.

Riess, Anita, ed. *Psychologie der Zahl.* Munich, 1973.

Schimmel, Annemarie. "Numbers." In *The Encyclopedia of Religion.* New York, 1987, vol. 11, pp. 13–19.

———. "Zahlensymbolik." In *Religion in Geschichte und Gegenwart.* 3d. ed. Tübingen, 1962, vol. 6, cols. 1861–1863.

Schulz, W. "Gesetze der Zahlenverschiebung im Mythos." *Mitteilungen der Anthropologischen Gesellschaft Wien*, 40 (1910).

Schwarz-Winkelhofer, Inge, and Hans Biedermann. *Das Buch der Zahlen und Symbole.* Munich, 1975.

Sergescu, Petre. *Histoire du nombre.* Paris, 1953.

Silver, Jules. *Deine Glückszahl. Eine moderne Zahlenmagie.* Geneva, 1976.

Taylor, Ariel Y. *Numerology Made Plain. The Science of Names and Numbers and the Law of Vibration.* 5th ed. Chicago, 1930.

Thimus, Albert Freiherr von. *Die harmonikale Symbolik des Alterthums.* Cologne, 1868. An extremely learned work that deals with "the esoteric number doctrines and harmony of the Pythagoreans in their relation to earlier Greek and oriental sources" in the first part and studies the cabalistic symbols of the *Sefer Yetsirah* in the second part.

Trevoux, Guy. *Lettres, chiffres et dieux.* Monaco, 1979.

Walton, Roy Page. *Names, Dates and Numbers. What they Mean to You.* New York, 1914.

Westcott, William Wym. *Numbers: Their Occult Powers and Mystic Virtues.* London, 1902.

Zimmermann, Werner. *Geheimnis der Zahlen: Zahl, Namen, Wesen, Schicksal.* Engelberg, 1948. Relation between number, name, character, and fate.

NUMBERS AND NUMERALS

Hartner, Willi. "Zahlen und Zahlsysteme bei Primitiv- und Hochkulturvölkern." *Paideuma* 2 (1941–1943), pp. 268–326. A first-class introduction to the various ways of expressing numbers in different cultures.

Ifrah, Georges. *Historie universelle des chiffres.* Paris, 1981. English translation by Lowell Bair, *From One to Zero.* New York, 1985.

Lipton, James. *An Exaltation of Larks.* New York, 1968. An amusing introduction into the expressions of counting various kinds of animals.

Menninger, Karl. *Zahlwort und Ziffer.* Göttingen, 1958. Translated by Paul Brioneer as *Number Words and Number Symbols.* Cambridge, Mass., 1969. A book with a great wealth of information written in a captivating style, about the various types of counting methods in the world. Excellent bibliography.

Schmidt, M. "Zahl und Zählen in Afrika." *Mitteilungen der Anthropologischen Gesellschaft Wien* 45 (1915), pp. 165–209.

ANCIENT NEAR EAST, INDIA, AND CHINA

Brunner-Traut, Emma. "Ägypten." *Enzyklopädie des Märchens*, Berlin, 1977.

Da Liu. *I Ching Numerology, Based on Shao Tung's Classic Plum Blossom Numerology*. San Francisco, 1979.

Eberhard, Wolfram. *Lexikon chinesischer Symbole*. Cologne, 1983. Contains contributions about number mysticism and individual numbers.

Glathe, A. "Die chinesischen Zahlen." *Mitteilungen der Deutschen Gesellschaft für Natur- und Völkerkunde Ostasiens* 36 (1932).

Granet, Marcel. *La Pensée chinoise*. Paris, 1934. Translated as *Das Chinesische Denken*. Munich, 1963.

Hopkins, Edward Washburn. "The Holy Numbers of the Rig Veda." *Oriental Studies* (1894), pp. 141–159.

———. "Numerical Formulae in the Veda and their Bearing on Vedic Criticism." *Journal of the American Oriental Society* 16 (1896).

Jeremias, Alfred. *Das Alte Testament im Lichte des Alten Orients*. 3d ed. Leipzig, 1916.

———. *Handbuch der altorientalischen Geisteskultur*. Leipzig, 1913.

Mahler, Eduard. "Das Himmelsjahr als Grundelement der altorientalischen Chronologie." *Zeitschrift der Deutschen Morgenländischen Gesellschaft* 60 (1906), pp. 825–838.

Sethe, Kurt. *Von Zahlen und Zahlworten bei den alten Ägyptern*. Strassbourg, 1916.

CLASSICAL AND POST-CLASSICAL THOUGHT

Arndt, O. "Zahlenmystik bei Philo." *Zeitschrift für Religionsgeschichte* 19 (1967), pp. 167–171.

Becker, O. *Das mathematische Denken der Antike*. Göttingen, 1957.

Delatte, Armand. *Etudes sur la littérature pythagoricienne*. Geneva, 1974.

Germain, G. *Homère et la mystique des nombres*. Paris, 1954.

Ghyka, Matila C. *Le Nombre d'or: rites et rythmes pythagoriciens dans le développement de la civilisation occidentale*. Paris, 1976.

Kérenyi, Karl. *Pythagoras und Orpheus*. 3rd ed. Zurich, 1950.

Ludwich, A. *Zahlsymbolik der griechischen Sakralbauten*. Königsberg, 1914.

Röd, W. *Geschichte der Philosophie*. Vol. 1, *Von Thales bis Demokrit*. Munich, 1976. The chapter about Pythagoras and the Pythagoreans is particularly valuable.

Staehler, Karl. *Die Zahlenmystik bei Philo von Alexandrien.* Leipzig, 1931.

Stenzel, Julius. *Zahl und Gestalt bei Plato und Aristoteles.* Leipzig, 1924.

Vasconcelos, J. *Pitágoras: una teoría del ritmo.* Mexico City, 1921.

BIBLICAL NUMBER MYSTICISM

Bond, Frederick Bligh. *Gematria: A preliminary investigation of the Cabala contained in the Coptic Gnostic books, and of a similar Gematria in the Greek text of the New Testament: Showing the presence of a system of teaching by means of the doctrinal significance of numbers, by which the holy names are clearly seen to represent aeonial relationships which can be conceived in a geometric sense and are capable of a typical expression of that order.* Reprint with new notes, London, 1977.

Fischer, Oskar. *Der Ursprung des Judentums im Lichte der alttestamentalischen Zahlensymbolik.* Leipzig, 1917.

——. *Orientalische und griechische Zahlen.* Leipzig, 1918. Attempts at connecting particular numbers and events. Thus, 13 is the "Yahwe factor" allegedly visible everywhere in the Bible.

——. *Auferstehungshoffnung in Zahlen.* Leipzig, 1920. Argues that the number 127 is the number of light and resurrection everywhere in the ancient Near East.

Friesenhahn, Peter. *Hellenistische Wortzahlensymbolik im Neuen Testament.* Leipzig, 1935. Reprint. Amsterdam, 1970. An extremely profound study of Hellenistic number mysticism in the New Testament.

Heller, Adolf Q. *Biblische Zahlensymbolik.* Reutlingen, 1936.

Kautzsch, E. "Zahlen bei den Hebräern," in RE 21, 598 ff.

Laubacher, E. *Phänomene der Zahl in der Bibel.* Neuewelt, 1955.

Weinreb, Friedrich. *Zahl, Zeichen, Wort. Das symbolische Universum der Bibelsprache.* Rowohlts deutsche Enzyklopädie, vol. 383. Reinbek bei Hamburg, 1978.

THE JEWISH TRADITION

Fredmann, Ruth Gruber. *The Passover Seder.* Philadelphia, 1981.

Müller, Ernst, ed. *Der Sohar. Das Heilige Buch der Kabbala.* Cologne, 1982.

"Numbers, Symbolic and Rhetorical Use." In: *Encyclopedia Judaica,* Jerusalem, 1971.

Reichstcin, H. *Praktisches Lehrbuch der Kabbala.* 5th ed. Berlin, 1954.

Scholem, Gerschom. *Zur Kabbala und ihrer Symbolik.* Zürich 1960, Frankfurt 1973.

Trachtenberg, Joshua. *Jewish Magic and Superstition.* New York, 1974.

Wünsche, August. "Die Zahlensprüche im Talmud und Midrasch." *Zeitschrift der Deutschen Morgenländischen Gesellschaft* 65 (1911), pp. 57–100, 396–421.

Zohar, The Book of Enlightenment. Translated and introduced by Daniel Chanan Matt. New York, 1983.

MEDIEVAL NUMBER ALLEGORESIS

Blüm, Hans Jürgen. *Der Altdeutsche Exodus. Strukturuntersuchungen zur Zahlenkomposition und Zahlensymbolik.* Amsterdam, 1974.

Brooks, D. *Number and Pattern in the Eighteenth-Century Novel.* Boston, 1973.

Butler, Christian. *Number Symbolism.* Ideas and Forms in English Literature 1. London, 1970. Numerous interesting and thought-provoking examples of number mysticism in general and in English medieval literature in particular.

Clark, Susan L. *The Poetics of Conversion: Number Symbolism and Alchemy in Gottfried's Tristan.* Bern, 1977.

Grossmann, Ursula. "Studien zur Zahlensymbolik des Frühmittelalters." *Zeitschrift für katholische Theologie* 76 (1954), pp. 19–54.

Hellgardt, E. *Zum Problem symbolbestimmter und formalästhetischer Zahlenkomposition in mittelalterlicher Literatur.* Munich, 1973.

Hopper, Vincent F. *Medieval Number Symbolism.* New York, 1938. An excellent, well-written introduction to medieval attitudes toward numbers, with special emphasis on Dante.

Knapptisch, A. *St. Augustins Zahlensymbolik.* Programm Graz, 1905.

Knopf, W. "Zur Geschichte der typischen Zahlen in der deutschen Literatur des Mittelalters." Ph.D. diss., University of Leipzig, 1902.

Kretschmar, P. *Mittelalterliche Zahlensymbolik und die Einteilung der Digesten- Vulgata.* Berlin, 1930.

Langesch, K. "Komposition und Zahlensymbolik in der mittelalterlichen Dichtung." In *Methode in Wissenschaft und Kunst des Mittelalters,* edited by Albert Zimmermann, pp. 106–151. Berlin, 1970.

Meyer, Heinz. *Die Zahlenallegorese im Mittelalter.* Munich, 1975. The most comprehensive study of its kind, offering a vast collection of materials with a thorough analysis of other studies in this field.

Peck, Russell A. "Number as Cosmic Language." In *Essays in the Numerical Criticisms of Medieval Literature*, edited by Caroline D. Eckhardt, pp. 15–64. Lewisburg, 1980. Survey of the Neoplatonic tradition of medieval number symbolism with analysis of the English visionary poem "The Pearl."

Rathofer, Johannes. "Structura Codicis -Ordo salutis. Zum Goldenen Evangelienbuch Heinrichs III." In *Miscellanea Medievalia* 16/2, pp. 333–355. Berlin and New York, 1984.

Reichmann, E. *Die Herrschaft der Zahl. Quantitatives Denken in der deutschen Aufklärung.* Stuttgart, 1968.

Schmitt, A. "Mathematik und Zahlenmystik Augustins." In *Festschrift der Görres-Gesellschaft,* pp. 533–566. Cologne, 1930.

Taeger, Burkhard. *Zahlensymbolik bei Hraban, bei Hincmar - und im "Heliand"? Studien zur Zahlensymbolik im Frühmittelalter.* Munich, 1970.

Thibaut de Langres. *Traité sur le symbolisme des nombres: un aspect de la mystique chrétienne au XIIe siècle.* Edited and translated by René Deleflie. Langres, 1978.

Tschirch, F. "Die Bedeutung der Rundzahl Hundert für den Umfang mittelalterlicher Dichtungen." In *Gestalt und Glaube, Festschrift für Oskar Söhngen.* pp. 77–88. Witten-Berlin, 1960.

Zimmermann, Albert, ed. *Mensura. Maß, Zahl, Zahlensymbolik im Mittelalter.* 2 vols. (= Miscellanea Mediaevalia, vols. 16/1–2.) Berlin and New York, 1983–1984.

ISLAM

Ahrens, Wilhelm. "Studien über die magischen Quadrate der Araber." *Der Islam* 7 (1917), pp. 186 ff.

Bergsträsser, Gotthelf. "Zu den magischen Quadraten." *Der Islam* 13 (1923), p. 227.

Birge, John K. *The Bektashi Order of Dervishes.* London, 1937. Reprint. London 1965.

Casanova, P. "Alphabetes magiques arabes." *Journal Asiatique* II 18 (1921), pp. 37–55; 19 (1922), pp. 250–262.

Corbin, Henry. "Le Livre du Glorieux de Jabir ibn Hayyan." *Eranos-Jahrbuch* 18 (1950), pp. 47–114.

Doerfer, Gerhard. *Die türkischen Elemente im Neupersischen.* Vol. 4. Wiesbaden, 1975.

Faddegon, J. M. "Note au sujet de l'aboujad." *Journal Asiatique* 220 (1932), pp. 139–148.

Fahd, Toufic. *La divination arabe.* Paris, 1970. A survey of Arabic divinatory methods.

Goldstein, B. R. "A Treatise on Number Theory from a Tenth-Century Arabic Source." *Centaurus* 10 (1964), pp. 129–260.

Goldziher, Ignaz. "Über Zahlenaberglauben im Islam." *Globus* 80 (1901), pp. 31–32.

———. "Über umschreibende Zahlenbezeichnungen in Arabischen." *Zeitschrift der Deutschen Morgenländischen Gesellschaft* 49 (1895), pp. 210–217.

Gölpǐnarlǐ, Abdulbaki. *Tasavvuftan dilimize geçen deyimler ve atasözleri.* Istanbul, 1977. Contains expressions connected with the magico-mystical use of certain numbers.

Hasluck, F. W. *Christianity and Islam under the Sultans.* Oxford, 1929. Important for the numbers 7, 12, and 40 in Byzantine and Turkish usage.

Horten, Max. *Die religiösen Vorstellungen des Volkes im heutigen Islam.* Halle 1917. Includes a still-valuable study of mystico-magical practices in popular Islam.

Huart, Clément. *Textes persans relatifs à la secte des Houroufis.* Leiden-London, 1909.

Kraus, Paul. *Jabir ibn Hayyan.* Cairo, 1943, vol. 2.

Kriss, Rudolf, and Hubert Kriss-Heinrich. *Volksglaube im Bereich des Islam.* Wiesbaden, 1950–1962. The second volume in particular contains considerable information on magic squares, numerology, and related customs.

Marquet, Yves. *La philosophie des Iḫwān al-Ṣafā.* Algiers, 1973. Includes a thorough study of numerology as found in the writings of the Brethren of Purity in tenth-century Basra.

Marzolph, Ulrich. *Die Vierzig Papageien.* Wiesbaden, 1983. Studies the concept of fortyfold repetition in narrative in the Islamic world, in particular in Iran.

———. *Typologie des persischen Volksmärchens.* Beirut-Wiesbaden, 1984. Contains numerous examples from Persian folktales, especially for 3, 7, and 40.

Mélikoff, Irène. "Nombres symboliques dans la littérature epico-religieuse des Turcs d'Anatolie." *Journal Asiatique* 250 (1962), pp. 435–446. Deals mainly with the role of 17 in popular Turkish works with a Shiite bias.

Nasr, Seyyed Hosseyn. *An Introduction to Islamic Cosmological Doctrines.* Cambridge, Mass., 1964.

———. *Science and Civilization in Islam.* Cambridge, Mass., 1968.

Önder, Mehmet, "Mevlevilikte 18." *Türk Folklor Araştırmaları* 85 (1956), p. 152. A brief study of the number 18 among the Mevlevis.

Winkler, Hans Alexander. *Siegel und Charaktere in der muhammadanischen Zauberei.* Berlin, 1930. Most important study of magic squares and related signs in Islam.

STUDIES ABOUT INDIVIDUAL NUMBERS

Two, Four, and Five

Backland, A. W. "Four, as a Sacred Number." *Journal of the Anthropological Institute* 25 (1895), pp. 95–101.

Buchmann, Monika. "Zwei + Zwie?" *Neue Zürcher Zeitung,* März 1984. A charming sketch concerning the twofold character of 2.

Geil, William Edgar. *The Sacred 5 of China.* Boston, 1926.

Knapp, Martin. *Pentagramma Veneris.* Basel, 1934. The discovery of the pentagram as the figure corresponding to the actual movement of the planet Venus.

Three

Abbot, A. E. *The Number Three: Its Occult Significance in Human Life.* London, 1963.

Brieger, Annemarie. "*Die urchristliche Trias Glaube, Hoffnung, Liebe.*" Ph.D. diss. University of Heidelberg, 1925.

Deonna, W. "Trois, superlatif absolu." *L'Antiquité Classique* 23 (1954), pp. 403–428.

Dumézil, Georges. *L'idéologie tripartite des Indo-Européens.* Brussels, 1958. Tries to show the tripartite structure of Indo-Germanic thought and activities.

Dundes, Alan. "The Number Three in American Culture." In *Every Man His Way.* Edited by Alan Dundes. Englewood Cliffs, NJ, 1968. A very interesting study about the predominance of tripartite structures in North American everyday life.

Erben, K. J. "O dvojici a o trojici v bájesloví slovanském." *Časopis musea královstvi českého* 31 (1857), pp. 268–286, 390–415. About the occurrence of two- and threefold structures in Slovak popular tradition.

Gerlitz, Peter. "Außerchristliche Einflüsse auf die Entwicklung des christlichen Trinitätsdogmas." Ph.D. diss., University of Marburg, 1960.

Göbel, Fritz. *Formen und Formeln der epischen Dreiheit in der griechischen Dichtung.* Stuttgart, 1935.

Goudy, H. *Trichotomy in Roman Law.* Oxford, 1910.

Guénon, René. *La Grande Triade.* Paris, 1957. Tries to explain the interaction of heaven, earth, and humans as well other triadic structures, mainly relying upon Chinese symbolism.

Günther, R. F. "Worauf beruht die Vorherrschaft der Drei im Menschen?" *Nord und Süd* 142 (1912), pp. 313–325.

Kaltenbrunner, Gerd-Klaus. "Magie der Zahl 3." *MUT* 237 (May 1987).

Kirfel, Willibald. *Die dreiköpfige Gottheit.* Bonn, 1948. A study of the phenomenon of tricephalic deities in India and elsewhere.

Lease, Emory B. "The Number Three, Mysterious, Mystic, Magic." *Classical Philology* 14 (1919), pp. 56–73.

Lehmann, A. *Dreiheit and dreifache Wiederholung im deutschen Volksmärchen.* Leipzig, 1914.

Lüthi, Max. "Die Dreizahl" in *Enzyklopädie des Märchens,* vol. III, Berlin and New York, 1981, cols. 851–868. A comprehensive analysis of the use of 3 in fairy tales and folktales.

Müller, Raimund. *Die Zahl Drei in Sage, Dichtung und Kunst.* Programm Teschen, 1903. A comprehensive study that deserves to be better known.

Nielsen, D. *Der dreieinige Gott in religionshistorischer Bedeutung.* 2 vols. Copenhagen, 1922, 1942. Study of the concept of divine triades in the history of religions.

Paine, Levi Leonard. *The Ethnic Trinities and Their Relations to the Christian Trinity.* Boston, 1901.

Philipp, Wolfgang. *Trinität ist unser Sein.* Hildesheim, 1983. A comprehensive study published 30 years after it was written. The Protestant theologian approaches the problem of trinity philosophically and reaches interesting conclusions.

Quecke, Kurt. "Die Lehre von den drei Körpern Buddhas, trikāyā." Ph.D. diss., University of Marburg, 1948.

Schöll, Hans-Christoph. *Die drei Ewigen.* Jena, 1936. A study of German village beliefs and traditions.

Seifert, Josef Leo. *Sinndeutung des Mythos: Die Trinität in den Mythen des Urvölker.* Vienna, 1954.

Söderblom, Nathan. *Vater, Sohn und Geist unter den heiligen Dreiheiten.*

Tübingen, 1909. One of the first theological studies dealing with trinity as a phenomenon known in various religious traditions.

Stade, B. "Die Dreizahl im Alten Testament." *Zeitschrift für alttestamentliche Wissenschaft*, 1906, pp. 124 ff.

Strand, T. A. *Tri-Ism, The Theory of the Trinity in Nature, Man and His Works*. New York, 1958.

Stuhlfauth, Georg. *Das Dreieck. Die Geschichte eines religiösen Symbols.* Stuttgart, 1937.

Tavamer, E. "Three as a Magic Number in Latin Literature." *Transactions of the American Philological Association* 47 (1916), pp. 117–143.

Usener, Heinrich. "Dreiheit." *Rheinisches Museum für Philologie* 58 (1903), pp. 1–47, 161–208, 321–362.

Seven

Adrian, F. von. "Die Siebenzahl im Geistesleben der Völker." *Mitteilungen der Anthropologischen Gesellschaft Wien*, 31 (1901), pp. 255 ff.

Bindel, Ernst. "Die Zahl Sieben in der Rosenkreuzerströmung." In *Die geistigen Grundlagen der Zahlen.* Stuttgart, 1958.

Boll, F. "Hebdomas." In Pauly-Wissowa, *Realenzyklopädie des klassischen Altertumes*, vol. 14/2, cols. 2547–2578, Stuttgart, 1893 ff.

Curtis, J. *A Dissertation upon Odd Numbers, particularly Number 7 and Number 9.* London, 1909.

Graf, J. H. *Die Zahl Sieben.* Basel, 1917.

Hartmann-Schmitz, Ursula. *Die Zahl Sieben im sunnitischen Islam. Studien anhand von Koran und Ḥadīt.* Göttingen, 1989.

Hehn, Johannes. *Siebenzahl und Sabbat bei den Babyloniern und im Alten Testament.* Leipzig, 1907.

———. "Zur Bedeutung der Siebenzahl." In *Abhandlungen zum Alten Testament. Festschrift für Karl Marti.* Gießen, 1925, pp. 128 ff.

Heiler, Friedrich. "Die Siebenzahl der Sakramente." In *Eine Heilige Kirche* 15 (1933), pp. 5–10.

Moin, Mohammad. *Taḥlīl-i Haft Paykar-i Niẓāmī.* Teheran, 1338sh/1959. A comprehensive study of the number 7 in Persian and Islamic tradition for a better understanding of Niẓāmī's epic *Haft Paykar* (The seven Beauties, or Princesses).

Röck, F. "Die Götter der sieben Planeten." *Anthropos* 14–15 (1919–1920).

Roscher, W. H. "Die Sieben- und Neunzahl in Kult und Mythologie der Griechen." *Abhandlungen der Sächsischen Gesellschaft der Wissenschaften*, phil. hist. Kl. 21/4. Leipzig, 1903.

————. *Die hippokratische Schrift von der Siebenzahl.* Paderborn, 1913.

Roux, Jean-Paul. "Les chiffres symboliques 7 et 9 chez les Turcs non musulmans." In *Revue de l'historie des religions* 168 (1965), pp. 29–53.

Seligman, S. "Das Siebenschläfer-Amulett." *Der Islam* 5 (1915), pp. 377–388.

Varley, Desmond. *Seven- the Number of Creation.* London, 1976. An attempt to see all of life in the rhythm of 7. An interesting collection of examples, but not convincing.

Zöckler, O. "Siebenzahl in der Bibel." In Albert Hauck, ed. *Real-Enzyklopädie für protestantische Theologie und Kirche,* Leipzig, 1896–1913, vol. 18, pp. 310 ff.

Eight

Schulz, W. "Das System der Acht im Lichte des Mythos." *Memnon* 4 (1910).

Underwood, P. A. "The Fountain of Life." *Dumbarton Oaks Papers* 5 (1950) Appendix B: "The Six and the Eight in Baptisteries and Fonts."

Nine

Kaegi, Adolf. "Die Neunzahl bei den Ostariern." *Philologische Abhandlungen für Schweizer-Sidler,* Zurich, 1891, pp. 50 ff.

Röck, F. "Neunmalneun und Siebenmal sieben im Alten Mexico." *Mitteilungen der Anthropologischen Gesellschaft Wien.* 1933.

Roscher, E. H. "Enneadische Studien." *Abhandlungen der Sächsischen Gesellschaft der Wissenschaften,* phil. hist. Kls. 26/1. Leipzig, 1907. Attempt at tracing the history of the 9 in Greek tradition.

Weinhold, Karl. "Die mystische Neunzahl bei den Deutschen." *Abhandlungen der Berliner Akademie der Wissenschaften,* phil. hist. Kls. 1897. An extremely thorough study about the role of 9 in Germanic lore.

Twelve and Thirteen

Böklen, Ernst. *Die "Unglückzahl" Dreizehn und ihre mythische Bedeutung.* Leipzig, 1913. A thorough study of the various applications of 13 and the possibility of exchanging 12 and 13 in many myths and folktales.

Graf, J. H. *Über Zahlenaberglauben, insbesondere die Zahl 13.* Bern, 1904.

Hoffmann, Paul. "Trisdekaidekaphobia Can Strike When You're Most Expecting It." Smithsonian Magazine, 1987, nr. pp. 123–128.

Mehlein, H. "Dreizehn" in *Reallexikon für Antike und Christentum*, ed. J. Th. Klauser. 4, cols. 313–333. Leipzig, 1941 ff.

Weinreich, Otto. *Lykische Zwölfgötterreliefs*. Heidelberg, 1913.

———. *Triskaidekadische Studien*. Gießen, 1916.

Forty and Larger Numbers

Berque, Jacques. "Les quarante." *Studia Orientalia* 1 (1956) pp. 215–216.

Borchardt, H. *55: der Schlüssel zur deutschen Sprache*. Stuttgart, 1974.

Burrows, E. "The number seventy in Semitic." *Orientalia*, NS 5 (1936, pp. 389–392.

Deny, Jean. "70–72 chez les Turcs." In *Mélanges Louis Massignon* I, Damascus, 1956, pp. 395–416. Mainly statistics showing how certain medieval Turkish writers center everything around 70.

Karahan, Abdulkadir. "Aperçu général sur les "Quarante ḥadīth" dans la littérature islamique." *Studia Islamica* 4 (1955), pp. 39–55.

———. *Türk-Islam edebiyatında Kırk hadis*. Istanbul, 1950. A fundamental study about the use of 40 in Turkish-Islamic traditions.

König, Eduard. "Die Zahl 40 und Verwandtes." *Zeitschrift der Deutschen Morgenländischen Gesellschaft* 61 (1907), pp. 913–917.

Rescher, Otto. "Einiges über die Zahl 40." *Zeitschrift der Deutschen Morgenländischen Gesellschaft* 65 (1911) pp. 517–520. A brief but very interesting article about popular uses of 40 in Islam.

———. "Einige nachträgliche Bemerkungen zur Zahl 40. im Arabischen, Türkischen und Persischen." *Der Islam* 4 (1913), pp. 157–159.

Roscher, W. H. "Die Zahl 40 im Glauben, Brauch und Schrifttum der Semiten." *Königlich Sächsische Gesellschaft der Wissenschaften*, phil. hist. Kl. 27, pp. 93–138. Leipzig, 1908.

———. "Die Tesserakonden und Tesserakondenlehren der Griechen und anderer Völker." *Berichte der Sächsischen Gesellschaft der Wissenschaften* 61, pp. 17 ff., Leipzig, 1919.

———. "Die Zahl 50 in Mythus, Kultus, Epos und Taktik der Hellenen und anderer Völker." *Abhandlungen der Königlich Sächsischen Akademie*, phil. hist. Kl. 33. Leipzig, 1917.

Rosenthal, Franz. "Nineteen." *Studia Biblica et Orientalia* 3 (1959), pp. 304–318.

Saintyves, P. *Deux mythes évangeliques. Les douze apôtres et les 72 disci-*

ples. Paris, 1938. About the application of the number 12 and 72 in early Christian tradition.

Steinschneider, Moriz. "Die kanonische Zahl der muhammedanischen Sekten und die Symbolik der Zahlen 70–73." *Zeitschrift der Deutschen Morgenländischen Gesellschaft* 4 (1856), pp. 145–170.

Weigand, H. J. "The 72 Languages of the World." *Germanic Review* 17 (1942), pp. 241–260.

ILLUSTRATION CREDITS

It is surprising that iconography in the area of number mysticism is still in its infancy. Hardly any of the numerous books on numerology deal extensively with the illustration of numerological principles in art history. Some medievalists have begun to work in this field, however; thus, Johannes Rathofer has recently shown that the *Golden Gospels* made for the German emperor Henry III is written and laid out according to symbolic numbers.

> What is the most difficult task? The one that seems to you
> easiest:
> To see with your eyes what is before your eyes.

This verse by Goethe, which can be applied very well to the mysterious relation between wisdom and number, is quoted at the beginning of the important study by the Swiss historian Wolfram von den Steinen: *Homo caelestis: The Word of the Arts in the Middle Ages* (Bern, 1965, 2 vols.). Some of the impressive pictorial examples from this work are reproduced here (plates 5–8) with the kind permission of the publisher, Francke-Verlag. We thank the Staatliche Museum in Berlin for the photograph of the Ishtar Gate from Babylon (p. 214).

We have tried to provide as many illustrations as possible. For the European Middle Ages and Renaissance we have consulted, among others, *Deutsches Leben der Vergangenheit in Bildern* (Jena, 1908, 2 vols.) and added some 1-page prints from the sixteenth and seventeenth centuries, based on the studies of Dorothy Alexander and Walter L. Strauss (New York, 1977), and of Wolfgang Harms (Munich, 1980).

For mysticism and the Baroque we relied on both original works and documentation such as *Geheimne Figuren der Rosenkreuzer* (Altona, 1785–1788), and the catalogue of the 1981 Athanasius Kirchner exhibition in Rastatt. Pictures about Eastern number symbolism are taken from Wolfgang Eberhard's *Lexikon chinesischer Symbole* (Cologne, 1983) and V.-F. Weber's *Koji Hú-ten* (Paris, 1923, 2 vols.), as well as the collective work *Das Licht des Ostens*, edited by Maximilian Kern (Leipzig, 1925) and Heinrich Zimmer's *Indische Mythen und Symbole* (Cologne, 1972).

The Maya examples rely upon Wolfgang Cordan's *Mexico* (Cologne, 1955) and the Ludwig Collection catalogue from the Wallraf-Richartz Museum, *Altamerika* (Cologne, 1974). The Islamic calligraphies come from the collection of Annemarie Schimmel; the cabalistic diagrams, from my own collection of Judaica.

We warmly thank the Belser Press for permission to publish the unique Leibniz Memorial Medal from G. W. Leibniz's *Zwei Briefe über das binäre Zahlsystem und die chinesische Philosophie* (Stuttgart, 1968).

Ulf Diederichs

INDEX

As the numbers, and in particular the integers between 1 and 0, occur throughout the chapters, they are not included in the index except for very special cases.

275, 276. *See also* Christ, Christian;
Gospel/s; Old Testament
Bidenthaler, Tobias (fl. 1620), 200
bija mantras, 125
binah, 17, 96, 133, 183
binary system, 3, 50, 259, 260
Binbir kilise (Anatolia), 278
Biruni, Abu Rayhan al- (d. 1048), 239
Bischoff, Erich, 35
bismillah, 49
Blocksberg, 173
Boethius (d. 524), 19
Boehme, Jakob (d. 1624), 70
Böklen, Ernst, 204
Bonaparte, Napoleon (d. 1821), 34
Bonn, 127
Book of the Dead, 224, 254
Book of Revelation, 5, 70, 86, 104, 124,
133, 137, 235, 276. *See also* John,
St., Revelation of
Bourbon, 34
brace (= 2), 56
Brahma, 44, 61, 90
Brethren of Purity in Basra, 18, 94,
119, 150, 167, 186, 189, 199, 211,
221
b'reshit, 49, 122
Brigid, Brigit, 66
Britain, 149
Browne, Sir Thomas (d. 1682), 107, 128
Buddha, 36, 68, 152, 159, 240, 269,
274; Buddhism, 68, 104, 152, 159,
160, 182, 223: *buddha, dharma,
samgha*, 68; Mahayana, 61;
Buddhist, 36, 52, 152, 272, 279
buduh, 30
Bukhari, Muhammad al- (d. 870), 147
Bungus (Bong), Petrus (16th cent.), 22,
164, 189, 277
Buni, al- (d. 1225), 34
Burger, Gottfried August (d. 1794), 240
Byzantium, Byzantine, 79, 90

Cabala, 5, 16–18, 42, 44, 53, 77, 96,
206, 222, 232, 241; cabalists,
cabalistic, 49, 52, 54, 65, 97, 116,
133, 182, 184, 185, 231, 232, 263,
264
cahier, 102
Cain, 133
calculus, 5

Calliope, 167
Calypso, 219
Canterbury, 122
Carthage, 20
Caspar, 79, 199
Cassiodorus (d. 585), 24
Cassirer, Ernst (d. 1945), 59
Catalan, 165
cat, 148, 237; nine lives, 173;
three-colored, 75
Catacombe of St. Priscilla, 158
Cathars, 266
Catholic, 45, 68, 69, 104, 135, 209,
246, 247, 277
Celts, Celtic, 4, 66, 73, 172, 173, 226,
241, 268
cempohuallis, 228
Central Asia/n, 167, 170, 266
centurio, 188, 271
chakras, 148
Chandogya Upanishad, 65, 216
Chang Ko-lao, 161
Chapman, George (d. 1634), 105
charbagh, 94; *Char Minar*, 102
Charlemagne (r. 768–814), 99, 274
Cherubim, Pl. 6
chiffre, cifra, 7
chihil, 249; Chihil Sutun, 249; *chilla*,
252, 253
China, 4, 29, 30, 68, 100, 104, 110,
112, 129, 132, 160, 161, 167, 169,
170, 197, 204, 243, 258, 266, 271,
278; Chinese, vii, 43, 50, 51, 88,
101, 110, 111, 144, 160, 169, 215,
223, 244, 259, 266, 267, 271, 277;
number system, 6
Chonsu, 61
Christ, 63, 64, 88, 90, 94, 107, 118, 126,
133, 135, 158, 159, 164, 184, 193,
196, 198, 203, 232, 238, 240, 242,
246, 247, 254, 266, 267; Christian,
11, 21, 42, 53, 56, 61, 66–68, 71, 74,
75, 78, 88, 96, 107, 118, 124, 126,
137, 142, 143, 147, 148, 152, 159,
178, 188, 193, 203, 204, 213, 221,
222, 232, 241, 242, 248, 249, 263,
269, 271, 276: exegesis, tradition, 54,
94, 133, 158, 164, 184, 193, 207, 229,
231, 237, 246, 250, 254, 264, 273;
Christianity, vii, 16, 19, 26, 44, 45,
63, 71, 109, 142, 158; disciples of, 273

Israel, 26, 36, 37, 132, 182, 193, 207,
213, 246; children of, 246, 254;
Israelites, 71, 108, 264
Istanbul, 154
Italy, Italian, 11, 99, 102, 124, 216

Jabir ibn Hayyan (9th cent.), 119, 220
Jacob, 65, 132, 133, 149, 205
Jain/s, 151, 269
Jain-Neubauer, Jutta, 176
Jair, judge, 240
jalal, jamal, 49
Janus, 10, 20, 53
Japan/ese, 61, 152, 163, 171, 226
jawahir al-khamsa, al- (Gwaliori), 114
Jehu, 254
Jenissey, 144
Jerome, St. (d. 420), 88
Jerusalem, 45; heavenly, 193
Jesuits, 236, 252
Jesus, 24, 53, 70, 71, 94, 119, 124, 149,
158, 184, 203, 222, 239–241, 248,
255, 266. *See also* Christ
Jews, 68, 108, 158, 244; Jewish, 11, 34,
36, 42, 52, 54, 56, 77, 128, 130, 139,
158, 159, 178, 182, 207, 222, 244,
263, 273; temple, 71
Joachim of Fiore (d. 1201), 73
Job, 184
John, St., 103, 135; Gospel of, 70, 272;
Revelation of, 86, 193, 273. *See also*
Book of Revelation
John XXII, pope (r. 1276–1277), 79
John of Sacrobosco (13th cent.), 7
John of Salisbury (d. 1180), 137
Jonah, 69
Jordan, 132, 193
Joseph, 133, 189
Joshua, 193
Jubal, 12
Judah, 246
Judaism, vii, 16, 44, 45, 71, 126, 130,
248, 255, 274. *See also* Jews, Jewish
Judas Iskhariot, 240
Judeo-Christian, 164, 199, 248
Junao, 20
Junayd (d. 910), Path of, 159
Jung, C. G. (d. 1961), 104, 106, 140
Jupiter, 20, 107, 110, 129, 142, 167,
168, 178, 197; square, 30, 31, Pl. 2;
square number, 32

Kaaba, 26, 146, 258
Kabir (15th cent.), 97
kaliyuga, 97
Kama, 97, 118, 176; *Kamasutra,* 259
Kandarpa. *See* Kama
Kanjur, 272
Käppeler, Bartholomäus (17th cent.), 137
Karbala, 186, 188, 266
Kékulé, August von (d. 1896), 126
Keller, Gottfried (d. 1890), 155
Kepler, Johannes (d. 1630), 24, 25, 107
Kerbholz, 5
keter, 17, 183
Kette (chain of partridge), 9
khams fi 'aynak, 116
khamsa, 27, 114
Khidr, 249
Khwarizmi, Muhammad ibn Musa
(d. ca. 845), 6, 7
Kindi, al- (d. after 870), 119
Kippumaki, Mt., 177
Kircher, Athanasius (d. 1680), 23, 167,
217, 232, 236, 265, Pl. 3
kirk (= 40), 249; *kirkayak,* 250;
kirklamak, 250; Kirklareli, 249
kiva, 99
Knapp, Dr. Martin, 107
Knoten, 9
Köbel (fl. 1539), 11
Koppel (string of horses), 9
Krishna, 176, 272
Kroton, 11
kun, 49; *kun fayakun,* 149
Kungfutse (d. 479 B.C.E.), 29, 111. *See
also* Confucius
Kurdistan, kurdish, 67, 221
Kutadgu bilig, 77, 170
Kwannon, 61
Kymrians, kymrish, 99, 172

lalah, 261
Lamech, 132, 133
Lao-tzu (ca. 600 B.C.E.), 60, 68, 170,
279
lata'if, 148
Latin, 6, 56, 79; America, 192, 236
Lau Ts'ai ho, 161
Law (Old Testament), 71; connected
with five, 37; connected with ten,
20, 21, 222, 232, 246, 273, 274
Leah and Rachel, 54, 132

Michael, 120
Middle East, 115, 244
Midgard, 59
Midianites, 275
Midrash, 264
Miegel, Agnes (d. 1964), 155
Milton, John (d. 1674), 23
Mina, 147
Minussian Tatars, 77
Mirror for Princes, 77
Mishna, 37
Mithras, 61, 143, 145, 239; Mithraic
    mysteries, 152, 156
Moenjo Daro, 101
Moghul India, 94, 171
moiras, 64
moksha, 97
Mongols, Mongolian, 77, 170, 201, 241
moon, 219; square, 31
Monreale (Sicily), Pl. 4
Mörike, Eduard (d. 1875), 118
Morley, S. G., 227
Moscow the Third Rome, 73
Moses, 23, 54, 71, 94, 107, 114, 119,
    133, 149, 182, 239, 246, 255, 263;
    Plato as "atticized Moses," 23
Most Beautiful Names of God, 241,
    270, 271
Mount of Olives, 79
mu'allaqat, 147
Mughira ibn Sa'id (d. 737), 220
Muhammad the Prophet, 14, 20, 67,
    82, 115, 119, 135, 147–149, 157,
    184, 186, 201, 211, 221, 239, 248–
    252, 263, 264, 270–272, 277
Muhammad ibn Isma'il (8th cent.), 186
Muharram, 186, 188
Müller, Raimund, 61
Münchhausen, Börries von (d. 1945),
    154
Munich, 66
muses, 166, 167, 174, 177
mushabbih, 67
Musibatnama ('Attar), 252
music, 11, 85, 112, 128, 135, 138, 139,
    235; Indian, 9, 85, 216; *musica
    coelestis*, 24
Muslim, 52, 68, 73, 78, 94, 96, 115,
    119, 133, 147, 157, 165, 168, 184,
    202, 220, 221, 224, 248, 251, 252,

266; India, 114, 157. *See also* Islam,
    Islamic
Mut, 61
*mysterium fascinans, mysterium
    tremendum*, 49

nandi, 92
Napoleon, 203. *See also* Bonaparte,
    Napoleon
Naramsin, 219
Nasir-i Khusraw (d. after 1072), 150
Nasrettin Hoca, 250
nasut, malakut, jabarut, lahut, 96
Nath Yogi, 269
naugaza, 175; *naulakha*, 175
nawruz, 146
Near East, 116, 126, 129, 130, 157,
    192, 193, 197, 203, 219, 224, 246,
    256, 266
Nebuchadnezzar II (r. 604–562 B.C.E.),
    214
Negenstärke, 174
nemontemi, 228
Neoplatonic, Platonism, 16, 18, 42, 44,
    88, 122, 178
Nepal, 187
Nereids, 256
Nergal, 209
neti neti, 42
netsah, 17, 183
Neunkräutersegen, 174; *neunäugig,
    neunhändig, neunmalklug,
    Neunmännerwerk*, 175
New Testament, 54, 109, 125, 133,
    207, 240. *See also* Gospel/s
New Year, 80, 112
Nicomachus of Gerasa (d. ca. 100), 19,
    20, 139
Nietzsche, Friedrich (d. 1900), 90, 155
Niflheim, 59
nine, 21, 27, 28, 30, 77, 164–179, 244;
    spheres, 18, 167
nineteen, 26, 224–225
ninety-nine, 269, 270
Niniveh, 213
nishanji, 100
Nizami (d. 1203), 106, 114, 148
Noah, 69, 119, 132, 149, 182, 219, 245
noon (= *none*), 164
Norden, Eduard, 69